Beyond The Lab Coats

A Journey Through Flavor, Formulas, and Life

José Barbosa and Yelena Barbosa

∞

Endless Thesis Holdings LLC

Published by Endless Thesis Holdings LLC
www.endlessthesis.com

ISBN (paperback): 979-8-9934544-0-5
ISBN (ebook): 979-8-9934544-1-2
ISBN (hardcover): 979-8-9934544-2-9

Library of Congress Control Number: 20259233898

Cover design: Urban Writers
Interior design and composition: Atticus

This is a work of nonfiction. Views expressed are those of the authors and do not represent the policies or positions of any current or former employer. Company and product names may be trademarks of their respective owners and are used for identification only. Some names and identifying details may have been changed.

Printed in the United States of America
First Edition

Disclaimer

The views and opinions expressed in this book are solely those of the authors and do not reflect the official policies or positions of any current or former employer, colleagues, or institutions mentioned. Company names are used to provide context for our personal experiences and are not intended to imply endorsement, sponsorship, or criticism.

This book is a memoir. While every effort has been made to portray events accurately, some names, identifying details, and timelines may have been changed to protect privacy or improve narrative flow. All anecdotes and insights are based on real experiences, but memory is fallible, especially when powered by caffeine and corporate deadlines.

This content is for storytelling and informational purposes only and should not be construed as professional, legal, nutritional, or medical advice. For any career, scientific, or dietary guidance, consult a licensed professional, preferably one who also enjoys a good metaphor.

Under no circumstances will any blame or legal responsibility be held against the publisher, or author, for any damages, reparation, or monetary loss due to the information contained within this book, either directly or indirectly.

Writing a book is like product development: It requires inspiration, patience, iteration, and an absurd number of snacks. This book would not have been possible without the countless people who shaped our journey, both personally and professionally.

First, to our families, your love and support gave us the courage to be vulnerable and the humor to survive chaos. To our four kids: You are our favorite experiment and our greatest joy. Thank you for letting us turn bedtime into productivity time.

To our mentors, colleagues, and fellow food scientists across the industry: Thank you for the lessons, debates, and shared late-night project scrambles. You helped sharpen our skills and gave us the stories worth telling. A special shoutout to the unsung heroes in QA, Regulatory, and Sensory, the true guardians of science and sanity.

To Lance and every person who gave us permission to share pieces of their journey, thank you for trusting us. Your stories brought this book to life.

To Pete, thank you for letting us tell your story with your name on it. Your generosity—and that simple yes—made this book braver.

To the readers who are discovering food science for the first time or rediscovering their place in it: We wrote this for you.

And finally, to each other, writing this book together reminded us that our favorite project is still "Team Barbosa." Whether in a lab, on a plane, or in the middle of bedtime negotiations, we are proud to build this life, and now this book, side by side.

Contents

Prologue

Throughout this memoir, we'll share stories and insights that reflect our family's philosophy and decision-making style. Since common sense, resilience, and consistent iterations aren't exactly page turners, we've had to summon all the charisma and charm we could muster to make these words sing mellifluously on paper. Writing gives us the rare chance to articulate what we often can't during real-life conversations; BBQs, birthday parties, and school pick-ups don't exactly lend themselves to TED-Ex™-worthy sermons. Between kids, distractions, and social etiquette, we usually get to share our beliefs in tweet-sized soundbites, enough to spark a thought, but not enough to unpack it. Plus, saying "hashtag anything" out loud is just so uncool, as our kids would say.

So, we did the next best thing: We wrote a book. A book that shows what it's like to live as a food science couple once the lab coats come off (and when they're on too). Our challenge? Make the words as digestible and joyful as a kid's first scoop of ice cream on a summer day. We want these pages to sing, to jump off the screen, even if you're reading them on the toilet while your morning coffee or probiotics are kicking in (no judgment!).

When we started our careers, we lived by a philosophy: *work hard, play hard*. We worked long hours, then stayed out even later. We pushed ourselves in the lab by day and squeezed every last juicy drop out of life by night. Sleep was optional.

Then came the suburbs of New England. Trading the pace of New York City for small New England towns felt like slamming on the brakes. By 7 p.m., the streets were quiet. The bars and cafés closed early. The nightlife... well, let's just say you had to make your own.

That philosophy we once thrived on slowly evolved. It became *work hard, then work harder*, not for the office, but for the family. Now, the late nights weren't for rooftop parties or last calls at the Brazen Fox bar in White Plains; they were for keeping the family machine running, prepping for the next day, and making sure one, and eventually four, little humans were fed, loved, and (eventually) asleep.

Still today, there is no sleep for us... just a different kind of night shift. Jim Gaffigan says it best: "*Every night before I get my one hour of sleep, I have the same thought: Well, that's a wrap on another day of acting like I know what I'm doing.*" That might as well be our mantra.

We used to chase deadlines and dance floors. Now we chase kids, carpools, and grocery lists. Turns out, *work hard, play hard* gradually became *work hard, love harder*, and maybe that's the most important philosophy we've ever lived by. Back then, we kept saying "one day": One day we will have a family. One day, we will be debt-free. One day... Somewhere along the way, we started saying "day one." And the first place that showed up wasn't in the boardroom; it was in our kitchen.

Food sits at the center of our lives, even though we eat to live and not live to eat. Because when a Puerto Rican food scientist who can't cook marries a Russian-born Jewish food scientist who cooks for fun, every dinner turns into a Cold War of flavor. Yet, somehow, we end up with perogies and rice and beans on the same plate with a side of matza ball soup and agree that the only thing missing is more Sazón™.

Our dinners are as varied as our adventures; Yelena and I are rarely static. Some might say we're allergic to routine (and interest charges); others just roll their eyes and ask, "Again?" We didn't settle near family or return to a familiar hometown. Trust us, we would have loved it! It just never fit the rhythm of our lives. Instead, we've become the couple that surprises everyone with the next move, the next plan, the next, "Wait, what are they doing now?" Our updates usually spark a "Shut the front door!" rather than a polite, "That's nice. So... how about them Yankees?"

Some psychologists might suggest we're running from ourselves. And to that, we kindly chuckle. In truth, it's the opposite. We're drawn toward something, not away. Without ever formally naming it, we've always lived by one unspoken rule: *take action*. Just start. Action creates momentum, and momentum took us from a fateful Mets game all the way to four kids, four states, and now, a book. That's how we've moved through life, whether it's building a family, changing careers, or writing this very book.

We don't always get it right. We've procrastinated like champs. We've watched entire TV series in denial of our to-do lists (ask us how many hours *Game of Thrones* stole from us). We don't claim to have it all figured out, but when something matters, we find a way to move forward. Not because we're fearless, but because we believe movement, any movement, is how growth happens.

Even after all that spent energy, we come back to our calendars, write down our goals, and recommit and recalibrate, usually over coffee and tea, during a very animated conversation in our beds, in the backyard, or in the car during the pickup routine.

We believe growth lives just outside the comfort zone, and though it's uncomfortable, it's worth it. Do we sometimes look at each other and quietly ask, "We'll be okay, right?" Absolutely. Then we get up and do the next thing. That, to us, is progress.

José: Many successful people describe life as signals and noise. They mention how they focus exclusively on signals. While I respect, relate to, and even admire that philosophy, our family does not live by signals alone; we've learned to fine-tune our radar. We choose our signals wisely. Family, purpose, integrity, and fun always make the list. Everything else? Just static we've learned to dial down.

I've often used the word "karma" at work to help others understand what I mean when I tell them to add positive energy to everything we do, whether we're having fun on projects or working off hours to finalize a last-minute request. Maybe karma is not some mystical force, but a well-worn equation: intention plus action equals outcome. Think of it as the compound interest of doing good.

Yelena: We've left roles and environments that didn't align with who we are. Not out of fear, but out of respect, for ourselves, for each other, and for the future we want to build for ourselves and our kids. We've tuned out the gossip, the drama, the unnecessary chaos, and instead put our energy into growth, laughter, and contribution to our family. Not because we're perfect—José, especially, is far from it—but because we know what brings us peace.

José: What? I'll ignore that. Dear reader, if you only read this prologue and put the book down, years from now, I hope you'll remember just one thing from this short read so far: The grind will test you. The world will pull at your attention with noise dressed up as opportunity. Yet the real art of growth is learning to hear your own signal through all of it and having the courage to follow it.

Yelena: If you ignore José's advice entirely, at least remember to wear comfortable shoes. You can't chase your signal if you've got blisters. When in doubt, check your compass. If it still points to kindness, curiosity, and joy, you've probably headed in the same direction we've embarked on. Come join us!

Introduction

We didn't set out to be food scientists. We didn't even set out to save the world, like most young people aspire to do. Even if we had, our capes would've been lost in the laundry before our first mission. The epitome of aimlessness, we entered the workforce with no purpose, but guess what? That's okay. That's the point. Purpose isn't something most people are born with. It's not waiting for you like a prize at the end of a scavenger hunt. It's something you stumble into, trip over, and eventually decide to keep. Maybe you've already found yours (congrats). Maybe you're still wandering, map upside down, wondering where it's hiding. Either way, the journey is the story. And that's where ours begins.

Somewhere along the way, we realized that we weren't chasing products, smiles, or even scientific breakthroughs. We were chasing something more elusive: purpose. It hides in the late nights, the product failures, the laughter between colleagues, and the most unexpected traditions. Beyond the lab coats, that's where purpose lives.

I'm José Barbosa. My career began with biology, pre-med, ambition, and a stubborn belief that what we put into the body can rewrite its story. I never planned to be here, yet here I am, standing in the gap between science and sustenance, chasing a vision that feels a little dangerous—dangerous because it dares to challenge the norm, dangerous because it requires walking into rooms where I'm not the loudest voice but still insist on being heard through the veil of kindness and empathetic melody, and dangerous because once you find purpose, you can't unsee it... and you can't go back to living a life that doesn't serve it.

I'm not chasing it alone. My better half is here with me. She didn't plan to become a food scientist either; she also only wanted to become a doctor. One summer internship at Cornell University rerouted her life. That path led to an engagement in Chicago, a wedding in Annapolis, a home in New Hampshire, four kids, and yes, now a book.

I'm Yelena Barbosa. I've spent 15 years in food science, most recently managing product development for a juice company in the Northeast. My work has taken me from flavor labs to factory floors, from early-morning brainstorms to midnight troubleshooting. While José has fought malnutrition from the supplement cabinet, I've made my mark in beverages, especially juices and frozen slushies, crafting drinks that refresh, delight, and sometimes surprise the people who drink them.

If you're intrigued by food science, wonderful, you'll find plenty of it here. But this book isn't just about our careers. It's about choices, risks, pivots, and small wins that shape a life. Our stories are the raw ingredients; the recipe is yours to write. Through the lens of food science, corporate adventures, parenting chaos, and the occasional "what were we thinking?" moments, we'll share what we've learned about navigating careers, relationships, and purpose. Maybe, just maybe, you'll see reflections of your own journey in ours.

We're not here to hand you a five-step plan or wrap life's messiness in a neat bow. We're here to tell you what it's really like: the risks that pay off, the ones that flop spectacularly, the mentors who change your trajectory, and the moments that make you question your sanity.

Along the way, you'll see why making a perfect emulsion isn't all that different from building a career (hint: both will separate if you ignore them too long). This isn't just a food science memoir; it's a story about career, family, leadership, and the messy pursuit of purpose, told through the lens of two food scientists.

Our desire is that when you reach the last page, you'll see what we see: that beyond the lab coats and the formulas, the real work is iterating on yourself until you've built a life and a purpose that's unshakable.

Part 1: Finding Our Flavor: The Journey Into Food Science

Chapter One

Setting the Table: Who We Are Before the Feast

Valet Parking

To the diner, every great meal restaurant seems effortless. You hand over your keys (if you are the valet type), take a breath, and step inside. Then, a plate is placed in front of you, polished and ready to enjoy. But what looks simple on the surface is almost always the result of hidden effort, resilience, and countless behind-the-scenes decisions.

This book is our way of inviting you into the kitchen. Before we serve the first plate, let us tell you who's cooking. We've spent decades in the kitchen of food science and leadership, and this book is our menu, crafted from the most delectable and sometimes messiest ingredients of our careers.

Yelena and I have been married for over a decade. We, and specifically our kids, are the ultimate blended experiment: one part Caribbean spice, one part Siberian frost, shaken vigorously in the centrifuge of corporate America and life as a family of six in 2025. Our kids still can't quite figure out whether to call their grandmothers *bubbe, babushka, grandma,* or a*buela.*

Yelena: One of my favorite things about our career paths is that we worked together twice, in the same company. We even worked on the same project, just from different angles. We'd brainstorm ideas at the dinner table, bounce solutions off each other in the car, and run experiments that benefited from two sets of expertise. Unlike us, it never gets old.

José: Throughout our combined 40+ years in the food and beverage and dietary supplement industries, 24 of them mine, we've learned that working together is both a hack and a flex, the exact opposite of what most people make it out to be. My own path has been shaped by earning a master's degree in food science and an MBA in Leadership, co-authoring a scientific publication on iron absorption, and leading at the executive level as the head of Research and Development and Regulatory. I've built and launched products across global brands, and the throughline in all of it is that the best work happens when the right people are in the room together.

For us, that sometimes means the same kitchen table. We know the look people give us when they hear, "I work with my spouse." They usually respond with, "I could never do that," followed by an oddly specific confession about what crime they'd commit to the other if they ever tried. We get it. Truly. For us, though, working side-by-side has always been a force multiplier. We miss it when it's gone. Maybe after finishing this book, we'll find ourselves building something together again.

My approach to food science and leadership has been shaped not just in the lab, but in the pages of the books I've devoured. I sometimes think of my career as a Tolstoy novel: sprawling, complex, full of unexpected characters who wander in and leave their mark. I've chased mysteries with Dan Brown-like urgency, searched for magic in the ordinary, like the way Garcia Márquez finds it in Macondo, and tried to strip away everything unnecessary in my writing and my work, Hemingway-style, until only the essential remains. Maybe that's why my formulations, like my stories, aren't just about what's in them, but about what's left out.

Yelena: If José's career feels like a Tolstoy novel, mine has always been closer to poetry: concise, deliberate, and obsessed with rhythm. I live for the precision of language and flavor, the way a haiku can hold a season or a sip can hold a memory. As José mentioned, we've worked at the same companies for years, but honestly, in those early days, it wasn't quite what people imagine when they hear "working with your spouse." We were on the same campus, yes, but on different teams and in different parts of the building, with our own projects, our own leadership, and our own professional identities. That separation gave us the space to have our own careers from 9 to 5, while still sharing the bigger experience together.

At PepsiCo—massive, bustling, and filled with dozens of developers in one location—we each marched in our own lane. We were both creating world-class beverages, but our paths didn't cross much during the workday. Occasionally, we'd have lunch together, but candidly, in those years, even sitting down for lunch at all was rare. That was the grind. We shared a belief, right or wrong, that to build your reputation, you had to go above and beyond, keep your head down, and throw yourself into the work.

So, when people say, "I could never work with my spouse," I smile, because in our 20s, we technically did, but not in the way they think. We didn't really "work together" until now, as we've started writing this book and building companies side by side. Like anyone who works closely with their partner, we've learned that communication, boundaries, and protecting work-life balance are key to staying happy, both at work and at home.

Back then, before kids, work consumed most of our lives, but it still felt special to carpool in together, drive home together, and debrief on our days. We understood each other's highs and lows without explanation because we were living the same culture and breathing the same air all day. Some things didn't even need to be said, especially on hard days.

Now, life is very different. We have a small tribe of kids, summer days filled with chaos and laughter, and new ambitions that demand just as much focus as any corporate job we've ever had. As we build these companies and write this book, we're figuring out a new cadence, one that lets us work, create, and chase our goals without missing the moments that matter most with our family. This is the real challenge of our lives right now: redefining what work-life balance looks like in this stage of life. I believe in us. I believe we can do it. I believe everyone can, too.

Where We've Hung Our Lab Coats

Between us, we've touched nearly every corner of the food, beverage, and supplement worlds, sometimes literally thanks to product spills.

Years	José Barbosa	Years	Yelena Barbosa
2001–2014	PepsiCo – Mastered beverage science, corporate psychology, and tooting my own horn.	2010–2014	PepsiCo – Mastered formulas, flavor, and the art of making safety goggles look stylish.
2014–2016	Keurig Green Mountain (now Keurig Dr. Pepper) – Rode the Kold wave before it crashed.	2014–2016	Keurig Green Mountain (now Keurig Dr. Pepper) – Liaison with Coke was an unexpected alliance.
2016–2020	Dutch State Mines (DSM) – Introduction to the Vitamin, Mineral, and Supplement Industry.	2016–2020	International Flavors and Fragrances (IFF); since merged with DuPont – Created flavor solutions that made even the most skeptical customers take notice.
2020–2025	FoodState Inc. (MegaFood) – 30 product launches in 22 months.	2021–2025	Welch's – I was grapeful for the experience of turning grapes into products I'm proud of.

Where we've hung our lab coats: a quick look at our career path

Why We're Telling These Stories

José: If resilience is the muscle of growth, then forgiveness is the water that keeps it alive. You can train, push, and endure as much as you want, but without forgiveness of yourself, of others, and of the past, you eventually run dry.

We've learned this the hard way. Life hands you plenty of advice, not all of it good, and sometimes from people who truly mean well. Parents, mentors, colleagues, even friends—many of them had the best intentions, but their words left us chasing paths we later had to unlearn. For a while, we carried resentment about that. Then we realized: Resentment is heavy. Forgiveness, on the other hand, is fuel.

Forgiveness doesn't excuse the mistake, and it doesn't rewrite history. It just loosens the grip the past has on you, making space for new growth. Like water to a plant, it's invisible once absorbed, but nothing survives without it.

We've forgiven many already, and, truthfully, there are still a few names on the list we wrestle with. That's part of the journey, too. What we want our kids to remember is this: Someday, you'll need to forgive. You'll need to forgive yourself for the choices that didn't turn out the way you hoped. You'll need to forgive others who weren't able to show up the way you wanted, and you'll need to forgive the past for being what it was.

Without forgiveness, the story stops. With it, the next chapter is always possible.

The truth is, there aren't many resources for aspiring food scientists, or even veterans of the field, that capture the strange mix of science, art, politics, and chaos that scientists navigate every day. Plenty of resources for leadership, but not too many for scientists, specifically food scientists.

As scientists, we're wired to hunt for answers, but rarely do we turn that same analytical lens on ourselves. Asking for help feels unnatural, as if vulnerability belongs in a petri dish rather than in conversation. Maybe it's because food science is the corporate world's best-kept secret, a profession with its own language that outsiders rarely understand. Or maybe we're too busy building shelf-stable products to sit down and tell our stories. When we do write, it's usually in the form of dense technical papers, layered with charts, statistics, and industry jargon, that require a PhD and a caffeine drip to fully absorb (don't blame us, blame the system).

It's like spending your day peering into a microscope to uncover hidden truths, yet never glancing up to see the beautiful mess of your own kitchen. That's where Yelena and I come in. This is our microscope turned outward: a collection of stories, lessons, and hard-earned laughs told through our combined autobiographical lenses. Consider it a (one-way) conversation over coffee or a vitamin-infused smoothie. We'll share our journey so you can chart your own, whether you're already in this field or just curious about the strange, delicious, and occasionally absurd world of food science.

In this new, skeptical, online world where characters hidden behind keyboards are somehow more trustworthy than experts, being right isn't enough for scientists.

Yuval Harari once said that it's far easier to build a compelling fiction than to find the truth. Facts are expensive to uncover, time-consuming to prove, and, let's be honest, often delivered in a way that would put a double espresso to sleep! Meanwhile, myths and half-truths come wrapped in vivid narratives, carried by storytellers who know how to light up the emotional parts of our brains. Once those stories take hold, the truth must fight uphill to catch up.

That's part of why food science is such an extraordinary career. It lives at the intersection of fact, creativity, and impact. It's also why we wrote this book. Whether you're working in the lab, leading a team, or just figuring out what's next in your life, you're constantly in your own Research and Development (R&D) cycle: testing, learning, tweaking, failing, improving. Purpose isn't a lightning bolt. It's an iteration. If we can tell our story in a way that makes you want to keep turning the pages, maybe it'll spark you to tell your own, and to follow it to a place where your work, your skills, and your happiness all line up.

That's our advantage as scientists, but it's also our opportunity. If we want science to guide decisions in boardrooms, kitchens, and policy debates, we must stop speaking only in data tables and peer-reviewed paragraphs. We must become storytellers. Not to embellish or distort, but to carry the truth with the same emotional power fiction enjoys.

This memoir is part of that experiment. If we can't tell our own story—the real, messy, factual one—in a way that makes you want to keep reading, then we're proving Harari right. However, if we can make you laugh (out loud, even better), lean in, and maybe even share one of these truths with someone else... then we've done our job, both as food scientists and as storytellers.

Since the best stories start with clarity, let's begin at the beginning: Lab Coats. Specifically, the science of food types. Many people ask us, "What exactly is food science?" I usually reply, "It's the world's most evanescent secret, a career path hiding in plain sight, like a fleeting susurrus in the corporate cacophony." Okay, don't roll your eyes at me, Yelena! Together, we answer with, "It's the world's best-kept career secret."

People imagine food scientists wearing hairnets all day, tasting ice cream, soda, or gummy bears... okay, sometimes it's exactly like that. In our careers, we've tasted some of the most delicious and most revolting things you could imagine, from chocolate-flavored protein drinks to lab-made plant proteins that smelled like boiled socks and left a lingering bitter taste, but only if you happened to drink water right after. Weird, but true!

More often than not, food science is about solving puzzles. Why is this emulsion separating? Why does this product taste metallic? Why did the market reject something that tested brilliantly with consumers?

Food science lives in the fascinating space where culinary art meets industrial-scale manufacturing. In our world, a delicious concept is just the starting point. The real craft is in making it safe, stable, legal, efficacious, great tasting, and consistent, whether you're producing 50 samples for a consumer test or 50 million bottles for store shelves.

If we've succeeded in getting your attention, lend us your ears (or should we say eyes?) and come along. You'll get two perspectives for the price of one, as both my wife and I unpack the worlds of food science, corporate America, and life itself, in a way that might just spark something in your own journey. Because behind every formula, every product launch, and every corporate adventure... there's a human story worth telling. This is ours.

Chapter Two

The First Bite: How We Found Food Science

M y grandmother's house always announced itself before you even stepped inside. The tall, ornate metal gates out front were heavy enough that you had to lean your whole weight into them to push them open. If they were locked, which was most of the time, the only way in was to call out, "*¡Abuela!*" until she came to unlatch them.

Her house sat right on the edge of a busy intersection. All day, trucks, cars, and the occasional blaring motorcycle wove an endless soundtrack. You'd think that kind of noise would be impossible to live with, but for us, it was the most comforting background music imaginable. It was the kind of steady, loud hum that faded into the walls, so much so that visitors were always surprised to learn we slept better because of it. To me, it was the sound of belonging.

Life, even now, can feel a lot like those gates and that noise. There's always something between you and where you want to be, a barrier to push through, a call for help to make, a constant, loud hum of distractions in the background. Yet, if you learn to work with it instead of against it, it can become part of your rhythm.

That rhythm carried us through long product launches, career pivots, raising four kids, and navigating an industry that never stops moving. We've learned that you don't have to wait for the noise to quiet down to make progress. You just have to keep showing up, calling "*¡Abuela!*" when you need a little help, and pushing those gates open.

For me, once the gates were open, there was always a table waiting inside, usually with a half-played game of dominoes. We'd spend hours around the table, slamming down tiles, laughing, and sometimes arguing about whether a play was actually legal. Domines was equal parts strategy and luck, much like a career. You wait, you plan, and then in one move, everything shifts.

Every career story has a first domino, that single moment when everything tilts in a new direction. For some, it's intentional, mapped out in color-coded planners years in advance. For me? Total accident. A happy one, but still an accident.

I didn't grow up dreaming of food science. I didn't even know it existed. My plan was to cure cancer, wear a white coat, and sign my name with "M.D." at the end. My parents loved that idea. My high school yearbook (go Pirates!) even voted me *Most Likely to Discover the Cure for Cancer*. And then... life laughed.

Fresh out of college with a biology degree and a stubborn streak, I was blasting résumés to anyone who would take them. In 2001, that meant scouring the classifieds in a printed newspaper, faxing applications from a machine that smelled faintly of ozone and cost 10 cents per page, and waiting for the landline to ring.

One listing caught my eye: *microbiology position*. I pictured myself stepping into a lab filled with rows of agar plates, the faint tang of ethanol in the air, incubators humming quietly in the background—a place where science smelled and sounded like science. I called, they scheduled me for an interview, and I was ready.

However, the morning I walked in, something felt off.

The building was spotless, but not in the sterile, overlit way of a lab. The floors were carpeted and the walls painted a soothing, corporate beige. Potted plants were arranged neatly along the walls. I passed cubicles instead of benches, printers instead of incubators. The air was cool and neutral, with no earthy whiff of growing microbes and no sharp hit of alcohol disinfectant. No faint "life" of a lab at all.

By the time I reached the receptionist's desk, my mental picture of agar plates and petri dishes was starting to flicker. This didn't look like a microbiology lab. This looked like... an office. A pristine, corporate office.

That's when I found out that it was, in fact, a temporary staffing agency. The realization hit me like a flat note in a favorite song. I hadn't come here to talk to a middleman; I wanted to be in the game, running experiments and making the calls myself. I was fresh out of college, fueled by ambition, and convinced I didn't need anyone to open doors for me. For a split second, I considered turning around and walking out.

The woman who interviewed me greeted me with a smile that was equal parts warmth and professionalism. She had that rare ability to make you feel comfortable while also sizing you up with sharp, unblinking precision. To this day, we still keep in touch (thanks to social media), though back then I didn't realize she'd be one of the first people to steer my career in an entirely unexpected direction.

She didn't waste any time with small talk.

"What do you want to do with your life?" she asked, leaning forward slightly, pen poised over her notepad. It wasn't a throwaway question; it was the kind that makes you sit up straighter.

I gave the only answer I'd rehearsed since I was a teenager, the one my parents loved to hear, the one that felt like the truest thing I knew about myself:

"I want to cure cancer."

I had even given the same answer to my Organic Chemistry professor when they asked why I was taking their class for the third time. Because I needed it. Because I needed to be a doctor. Because I needed to cure cancer.

A few days later, the interviewer called back with an eight-day assignment at PepsiCo washing beakers. I politely declined.

The next day, she called again: "It's just eight days. Take the money, and we'll keep looking for you."

Fine. I relented. Just like that, my accidental entry into food science began.

When I walked into that building, it was like stepping into Willy Wonka's factory, but for beverages: colors, aromas, fizzing carbonation, shiny stainless-steel tanks.

My very first words to my new manager were, "Wait, you mean people get paid to make and drink soda?"

Eight days turned into sixteen. Sixteen turned into a full-time job. A full-time job turned into a career developing products for iconic brands like Mirinda, Pepsi, and Starbucks. Somewhere along the way, it also turned into meeting my future wife.

Yelena: If you had asked me at 18 whether I saw myself in food science, I would have said no. I had my heart set on becoming a doctor, too. But life, as it often does, had other plans. My journey took me from chemistry at Stony Brook University, to an internship at Cornell, to a master's in food science at The Ohio State University, to PepsiCo, where—spoiler alert, I met José. For me, food science wasn't an accident; it was a pivot. That pivot ended up being the most rewarding twist of my career (and my life).

Both: The beauty of this field is that no two paths into it are alike. Some stumble in like José. Others take a calculated turn, like Yelena. Either way, the first bite, the appetizer, hooks you. Once you've tasted it, there's no going back.

Anything Else Before I Take Your Order?

José: I felt guilty taking a paycheck in those early years, as the job, to me, was a privilege and didn't feel like "work" at all. I developed products for iconic international brands, joined employee resource groups (ERGs), and played in company intramural sports. I had found my niche; I had found my community.

That community turned into a team, and that team became the gold standard. We didn't just hit targets; we scored the highest employee engagement scores the company had seen in that era. It wasn't by accident either. The company even studied us, trying to figure out what made us tick. Why did this group of scientists and managers thrive when so many others fizzled? How did we do it? What was the secret sauce? What's the recipe, or should I call it formulation, for a high-performing team?

The truth is, we had stumbled into something rare—a mix of trust, humor, curiosity, and a complete absence of ego. Trust ran high because we all backed each other without expecting kickbacks or rewards. Why take on more work and risk not finishing your own? Because every single one of us would do it for the other, no questions asked. That kind of trust is rare, and once you've experienced it, you spend the rest of your career chasing it. So, like any good scientists, we reverse-engineered the formula. What was the secret sauce? Okay, I'll share it early in the book. It's easy to remember:

The Formula for a High-Performing Team

(serves unlimited, keeps indefinitely)

- **1 part "positive manager(s)"**: the kind who trust you, challenge you, and don't hover over your desk or beaker.

- **2 parts "no ego"**: leave the self-importance at the door; this dish doesn't need it.

- **3 parts "curiosity-filled mindset"**: because the best teams are fueled by "What if...?"

- **Garnish generously with a sense of humor**: sprinkle often; it keeps the whole thing from boiling over.

Pro tip: Serve daily and never let it go stale.

Now, if you're curious about food science but perfectly happy to leave the sausage-making to the scientists, that's fine too. This book won't hand you a diploma or a framed certificate. But it will give you something rarer: a behind-the-scenes pass into the world of food science. You'll see what really happens from idea to grocery shelf: the experiments that work, the ones that go sideways, the "What were we thinking?" moments, and the small, quiet wins that make it all worthwhile.

This memoir isn't just about food science. It's about the people who make it happen: the leaders, the managers, the mentors, and yes, the occasional micromanager who could turn a team meeting into an Olympic sport. It's about the choices we make under pressure, the pivots when plans fall apart, and the humor that keeps us from losing our minds along the way.

If you've ever wanted a peek behind the curtain, welcome. Grab your lab coat, or your leadership hat... or both. We promise to keep the jargon digestible, the lessons useful, and the stories seasoned with enough spice to keep you coming back for seconds.

Every good meal starts with an appetizer, so let's begin with ours: the messy, winding, often accidental way we found ourselves in food science in the first place, and why that path, however unplanned, set the table for everything that came after. The tour is about to begin. The table's set, the pans are sizzling, and the conversation's just beginning. Safety goggles are optional... but strongly encouraged.

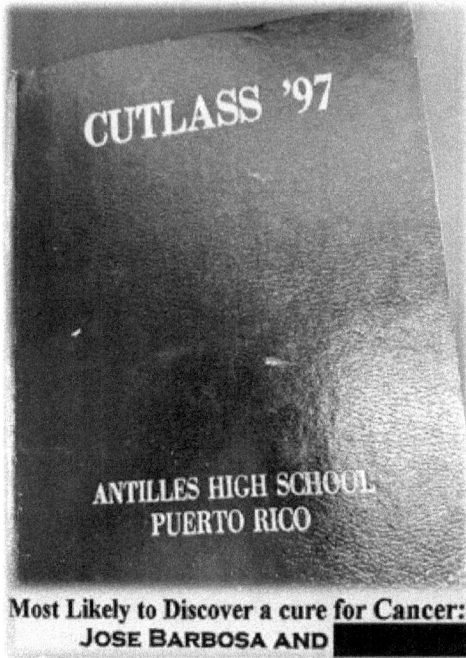

Figure 1. "Apparently, my classmates thought I'd cure cancer. I'm still working on it—just in supplement form."

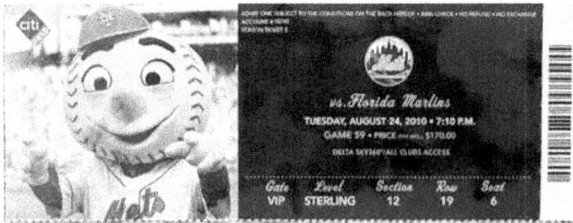

Figure 2. "The day science met serendipity—and baseball got the assist."

Chapter Three

Acquired Tastes: Choosing Food Science

José: You know how most people have a "career plan"? Something tidy like:

   1. Pick a major that makes sense.

   2. Land a job that matches said major.

   3. Work hard, get promoted, and make your parents proud.

Yeah... my path was more like:

   1. Wander into a major because it sounded interesting.

   2. Realize halfway through that "interesting" comes with organic chemistry.

   3. Stick with it anyway, partly out of stubbornness, partly

because it was starting to feel like the secret menu of careers.

Choosing food science wasn't exactly a lightning bolt moment. It was more like spotting an intriguing appetizer at a party, thinking, *Sure, I'll try that,* and then suddenly realizing you've made it your whole meal!

You can already tell that my "plan" looked more like an episode of *Chopped,* that cooking show where chefs get a basket of random ingredients and have to create a dish. My basket was a biology degree, a half-finished pre-med track, a temporary job washing beakers, and the challenge of turning it all into a 24-year career in food science. Just like the chefs on *Chopped,* your best strategy is to work with the ingredients in the basket, rather than trying to cook the dish you imagined before opening the basket.

For instance, when I arrived at Manhattanville College (now Manhattanville University), I didn't fully understand why a preceptorial curriculum required philosophy classes. I came to study science; why would I need Kant when I had chemistry? Somewhere between lectures and late-night study sessions, I began to see it. In Kant's work, I found a concept that resonated deeply: the autonomy of the learner, the idea that we hold agency over our own education and our own growth. It wasn't about memorizing dates and doctrines; it was about shaping the way one thinks, questions, and decides.

This is the part where I tip my hat to the liberal arts model, the one that some dismiss as impractical. For me, it was the opposite. It forced me out of the comfortable confines of formulas and lab reports, challenged me to wrestle with ideas bigger than my major, and helped me develop the kind of critical thinking that would later be just as important in the boardroom as it was in the lab.

I wouldn't have even stepped foot on that campus if one person hadn't believed in me: the Director of Admissions, who coincidentally was also from Puerto Rico. He saw potential in me before I fully saw it myself. He recruited me from the island and encouraged me to take the leap to study in what, for me, was a "foreign" land. It's no surprise he later became one of my first mentors. In those early days, when the weight of academic and cultural adjustment felt overwhelming, his guidance and reassurance were a lifeline.

Manhattanville didn't just give me an education; it gave me a foundation. It was a place where I learned that knowledge is more than information, that growth is more than grades, and that sometimes the most valuable lessons come from the classes you never thought you needed.

Yelena's path was more strategic than mine, though she'll admit there were still a few surprise ingredients along the way. She had her eyes on a medical career, but found her love for food science through hands-on internships and research opportunities. While I was stumbling into my career like a contestant who didn't read the challenge card, Yelena was out there plating her future like a Michelin-star chef.

Here's where it gets interesting:

Do you know how many food scientists originally wanted to be doctors? Yeah... I don't know either. But I do know that in our household, the answer is two out of two. Which, if you're a data-driven person, clearly means 100% of food scientists wanted to be doctors first.

Now, I have proof it's not just us. I recently learned that Vickie Kloeris, author of *Space Bites: Reflections of a NASA scientist*, also took pre-med courses and once planned to trade in a stethoscope rather than a lab coat. Apparently, there's a cosmic link between wanting to heal people and wanting to feed them (or, in NASA's case, feed them in space). I'm starting to think it's an unspoken rite of passage:

Step 1: Dream of being a doctor.

Step 2: Realize there are other ways to make an impact.

Step 3: Trade in the white coat for a hairnet, safety goggles, and the occasional food-grade spatula.

The common thread? We all wanted to help people. Turns out, you can still save lives; you're just doing it one shelf-stable beverage or nutrient-dense snack at a time. (Relax. This isn't product labeling. No nutrient content claim substantiation. No FDA warning letters. Just a humble book trying to make "nutrient-dense" cool again.)

The beauty of food science is that there's no single way in. You can be the accidental hire who never leaves (me), the focused student who discovers the field early (Yelena), or someone pivoting from another industry entirely. No matter how you arrive, you'll eventually find yourself answering the same question over and over: "So, what exactly do you do?"

That, dear reader, is where the fun begins.

Growing up, my culinary range was... limited. We're talking pizza, chicken (without skin), and plain rice, on repeat, well into high school. Like many kids of the late '80s and early '90s, I took my daily dose of Flintstones™ vitamins. While my diet might've suggested I was teetering on the edge of malnourishment, those chalky, chewable little dinosaurs made sure I was, at worst, only nutritionally deficient. Still, something about those vitamins must've planted a seed in my psyche, or maybe it was life's version of foreshadowing.

If we had a penny for every time someone asked, "So, what exactly is food science?" we'd have about a dollar, but that's not the point. According to the Institute of Food Technologists (IFT), food science is "the study of the physical, biological, and chemical makeup of food; the causes of food deterioration; and the concepts underlying food processing." In simpler terms, we're the folks figuring out how to make your food taste good, stay fresh, and not send you to the hospital. It's a field that touches everything from culinary innovation and snacks to supplements, confections, pharmaceuticals, and a few others we'll touch on throughout the book.

Food scientists wear many hats, sometimes even literal hairnets. We've known product developers who became project managers, formulators who pivoted into sales, and R&D leaders who ended up in the C-Suite. It's a surprisingly versatile career path hiding in plain sight.

Still, confusion persists. People often mix us up with chefs, dietitians, and even bartenders. And we get questions—oh, do we get questions:

- "Should I eat ice cream every night for dinner?"

- "Is soda gluten-free?"

- "Can I take ashwagandha with my meds?"

While we may know a thing or two, we're not your doctor or your nutritionist, so we usually answer with a mix of science, experience, and a healthy dose of "it depends." The most common question is, "Should we eat this food or drink that drink?" Yelena and I have become accustomed to giving the same answer over and over: "Everything in moderation."

The Case for Moderation

You've probably heard the phrase "everything in moderation" so many times that it's basically elevator music. We get it—it can sound vague, like a polite way of saying, "I don't want to explain myself." For us, it's not a dodge. It's a guiding principle we live by (even when it's inconvenient or competing with the gravitational pull of StarCraft 2™).

We'll be the first to admit: Nutrition is complex. Your needs are not your neighbor's needs. If we tried to unpack the biochemical, emotional, and cultural intricacies of how and why we eat, we'd need a whole other book, and probably a podcast series with guest experts and theme music. However, complexity doesn't mean we need to throw away simple truths. "Everything in moderation" is one of those truths that sticks because it works, if you give it a chance and stop rolling your eyes.

In our house, moderation is practiced in real time with four kids ranging from toddler to tween, ages two through ten. That age spread means our definition of balance shifts with circumstance. Yes, we let them eat sweets. No, they don't live off cotton candy and chicken nuggets. Yes, we allow screens. No, we don't think bedtime should be delayed for "just one more level." We give them (and ourselves) room to indulge, but we also know when it's time to pivot. If a travel day means more iPad time, a normal week might mean a quiet, no-fanfare tech detox.

We've learned that moderation isn't about perfection. It's about noticing the drift, forgiving yourself, and nudging things back into balance. Like realizing you've just spent three hours building Lego™ structures and deciding it's probably time to get some fresh air, or at least eat something that isn't made of plastic bricks. Sometimes it's letting the kids have ice cream before dinner because it's the first warm day of spring, and then making sure dinner still includes something green.

So, if moderation feels too soft or wishy-washy, give it a second look. It's not a loophole. It's a philosophy, and around here, it's kept us grounded, healthy(ish), and more or less sane.

Falling Into Food Science

To be frank—although, as you know by the introduction, I'm not Frank, I'm José, but I digress—I'm not sure many people choose to be food scientists by free will and exposure to the field. Many of us fall into this field by happenstance. It's an easier path than medicine, though when done right, it can be just as lucrative. Yes, you can make upwards of half a million dollars in total compensation, i.e., salary and bonus, in food science roles! Granted, those salaries are typically found in the highest-level roles, but it's still an achievable figure within the field. And I'm not joking. If I were, I'd say something like, "I wanted to write a joke about Sodium, but I was like, Na, people won't understand". Okay, let's be serious now. Choosing a career in food science means you love science. Unless you're one of the lucky few who learned about it in school, your gateway is probably through biochemistry, biology, chemistry, organic chemistry, or even microbiology. All roads can lead here.

Here's how we found ours.

Yelena's Leap

Yelena: I started out a PepsiCo as an analytical chemist because, well, that's what I did in grad school: analyzing tomatoes to make beverages for cancer patients. Super niche skill, right? I knew absolutely nothing about product development, but once I got to Pepsi and saw the product developers strutting around creating fun drinks, I thought, *Hey, why am I just measuring stuff when I could be making slushies?*

José: I'll interject here—if only she had spoken to a developer before she joined, right?

Yelena: Yes, if only... now back to my story. It took a year of dropping hints to my managers, a sprinkle of patience, and a dash of luck before an opening popped up on the food service team. Next thing I knew, I was developing drinks for Taco Bell™ and 7-Eleven™, crafting slushies and fountain beverages, and officially swapping my lab goggles for taste-testing cups galore.

A Twist of Fate

Yelena: The story of how I chose PepsiCo is worth sharing for the romantics and those who believe in fate.

One day, I was in an ice cream shop at The Ohio State University (go Buckeyes!) when a company I'd applied to called back. I'd sent out a stack of applications, and this was the first nibble. The role was in the Midwest. I loved OSU, but I wanted the East Coast. Still, with no other offers, I accepted.

I'd already had engaging conversations with PepsiCo. I'd even spoken to a real product developer there who explained what they did compared to analytical chemists. He seemed nice and chatty. However, no offer had come, and the Midwest company was in hand. I took it, with all the enthusiasm of someone accepting a consolation prize. Something about that conversation had destiny written all over it, though, and something inside was pulling me toward it.

Three days later, the offer I really wanted came in! I rescinded my earlier acceptance and jumped for joy. Gut feeling: 1. Playing it safe: 0. I am amazed at my courage, the risk I took, and the implications for my career of calling the original company back and rescinding my acceptance so I could pursue my true joy. People say, "Your gut feeling is usually correct." At that moment, I couldn't disagree. Fate had a lot more than courage in store for me after that decision.

José: The chatty developer? That was me.

Proof in the Emails

From: [Recruiter]
  To: Barbosa, José L.
  Cc: Poklad, Yelena
  Hi José,
  If you have some spare time away from sampling emulsions, I'd love you to talk with Yelena Poklad (pronounced Yell-EE-na). She'd like to chat about [Company Name] R&D. She's got a BS in Chemistry from Stony Brook and an MS in Food Science from the Ohio State U.

José: *Hey [Recruiter], sure thing. Yelena, happy to hear you're interested! I'm a product developer here and love it. When's a good time to chat? Are you free today?*

Yelena: *Thanks for taking the time! This afternoon works. I look forward to talking with you.*

Just like that... love at first sip.

José's First Taste

Ice wine can't be rushed. Neither can a career, a family, nor a purpose. All require frost, waiting, and the courage to harvest when others wouldn't.

I'd already been at PepsiCo for nine years when I spoke to that lovely young lady. I started as the lowest-level technician. Cubicles were for scientists; I sat outside in the lab. I wanted in. Within two years, I had my own cubicle, surrounded by colleagues who welcomed me without ego or condescension—not to be confused with condensation.

The lab was magical! Difficult science became a game of mixing ingredients and tasting delicious beverages. Taught to add flavor first before any other ingredient, I quickly discovered that adding expensive flavor at the start was risky and wasteful, like garnishing a stew you hadn't tasted yet. Serendipitously, it also mirrored manufacturing best practices. Within a week, I had already changed the best practices in the lab. Was it luck? Karma? I believe it was timing and curiosity, two key ingredients to this topsy-turvy world of ours. To lessons for life: Timing is everything, and bring your own curiosity.

My own path started with a curiosity for how flavors worked and how things we eat could be engineered to taste better, last longer, or be more nutritious. That curiosity eventually landed me in product development, where, to my delight, I learned that "tasting things at work" was part of the job description. From there, well, you're holding the rest of the story in your hands.

Yelena's entry point into this career was a little different. As she stated earlier, she began her career as an analytical chemist, laser-focused on the tiny details most people never think about. She wasn't setting out to create the next blockbuster snack; she was solving molecular puzzles. It turns out those puzzles had a lot to do with food, and before long, she was pulled into the same strange, delicious orbit as me. We didn't just find food science; we collided with it! Once it hooked us, we never left.

To this day, random food-science facts live rent-free in my head. Benzaldehyde tastes like cherry... or almond, depending on the beverage's pH. Sugar goes through an inversion process when exposed to heat or just left alone over time. The conversion factor for inverted sugar—yes, I know it. The hardness of water can change the color and flavor profile of your beverage—yes, I know that too. Don't ask why I still know that, or the many other random facts we all pick along the way. I'm sure this is true in any profession; ours just happens to be food science-centric.

Disclaimer Time

We may sound like experts, and we are, but we're not doctors. We can't diagnose you, prescribe meds, or create a personalized diet plan. Nutrition is deeply personal and nuanced. What we can do is share our stories and insights.

Speaking of which...

The Night I Played Doctor

I have come alarmingly close to playing doctor, like the time I delivered our fourth child, a baby girl, on the side of Interstate 93 Northbound between Derry and Windham, New Hampshire, on a cold and rainy April night. It was the last night of April, to be exact. Yes, really. Weird, but true.

It's one of our favorite icebreaker stories, and a surprisingly popular one at dinner parties. Here's how it went down:

José: Knowing our family was about to outgrow your standard five-seater, we brought a seven-passenger car. I even joked, "Just don't have the baby in the new car," and thought we had a mutual understanding. The night Yelena's water broke, we assumed we had plenty of time; after all, we had previous experience with three other deliveries and thought we were pros. The hospital was 30 minutes away. Our amazing neighbor arrived within five minutes, despite the pouring rain and the fact that it was nearly 11:00 p.m.

Yelena: For the record, I knew we didn't have plenty of time. This was our fourth baby, but there's only so much you can argue when you're trying not to scream in the passenger seat. I remember saying, "Call 911. This baby is coming." When José gave me that calm, "We'll be there soon" energy, I repeated myself, but this time with much less calm and much more rage.

José: She wasn't kidding. The dispatcher told me to pull over and check if the baby was crowning. I was just trying to prove her wrong, until I looked over and realized that she was very, very much right. The baby was indeed crowning!

Yelena: You're welcome.

José: I ran to the other side of the car as the dispatcher warned me, "It'll be slippery, like catching a fish." Well, just as I opened the door, the baby literally popped out. No joke. She launched into my arms, backside up, and started crying immediately—which, if you're wondering, is exactly the sound you want to hear.

Yelena: Meanwhile, I was sitting in the passenger side front seat... *still having contractions*! For those who may not know, the placenta still needed to come out, and it hadn't yet. Delivering a placenta on the side of the highway comes with a real risk of hemorrhaging, so I had to sit, wait, and keep contracting in that seat until the ambulance arrived and got us to the hospital. It was like having a second labor, with a baby in my arms.

José: It was chaos. The dispatcher started asking me to grab a shoelace, for what, I can only guess. Just as I was scanning the car for anything remotely helpful, there was a knock on the driver's side window. The ambulance had arrived. The EMTs sprang into action and earned their Stork Pins that night. As for me, well, I'm still waiting for mine. (If anyone from the NH Department of Heroic Deliveries is reading this, please call me.)

Yelena: They did indeed earn their Stork Pins that night. I told them they should give one to José, too, or did I? Maybe I asked for painkillers; it's all a blur.

José: Yes, we had the car professionally detailed the next day... by the prestigious Barbosa Detailing Company. Staffed by one very surprised, very tired, and slightly traumatized dad. Oh, and somewhere in all the excitement, I lost a sweater. No idea how or where. Maybe it got drafted into emergency blanket duty. Maybe it's living a new life somewhere off Exit 3. Either way, it's gone.

Life Before the Lab

We joke that every great product launch is like giving birth: stressful, messy, and occasionally unpredictable. But that night reminded us how laughably unprepared you can still be, even when you think you know the drill.

It was raw, unpolished life—the kind you can't rehearse, can't optimize, and can't run a post-mortem on in a PowerPoint deck. Yet, somehow, we got through it together, running on instinct, adrenaline, and just enough humor to keep from completely unraveling.

In corporate life, you're taught to control what you can, anticipate every risk, and have a Plan B, C, and D. It's a survival skill in case budgets get unexpectedly cut, markets tank like a toddler denied dessert, or projects get cut like a bad haircut. How does that translate to life? Arnold Schwarzenegger might be on to something when he says having a Plan B can distract you from Plan A. Purpose doesn't have an "almost." It's either yours or it isn't.

Sometimes, you're standing in the rain on the side of I-93 with a newborn in your arms, and your only plan is: "Don't drop the baby." That night didn't just add another child to our family; it added a reminder we carry into everything from parenting to product launches: Even when you think you've got it under control, life will hand you a wild card. If you're clear on your purpose, you'll know exactly which hand to play, no backup plan required.

Chapter Four

The Kitchen Prep: Early Career Grind

Navigating the Job Hunt With Humor, Heart, and the Occasional Emotional Breakdown

B ack in the lab, the surprises were a little less dramatic: no highway deliveries or shoelace requests from 911 dispatchers, but the lessons still applied. The early years in food science had their own kind of chaos, their own unexpected curveballs that kept you humble. Just like that night on I-93, the only way forward was to stay calm, adapt fast, and trust people around you.

If you're a student, a recent graduate, or someone mid-career wondering what's next, allow us to say this upfront: The job hunt will test your resilience. It's part matchmaking, part endurance sport, and part existential crisis. But if you do it right, it's also fun.

Yelena and I have both walked this twisted path, sometimes together, sometimes separately, but always with snacks. We've filled out endless job applications (each with its own login system, of course), survived multi-hour behavioral tests that felt like pop psychology marathons, and sat through interviews that ranged from exhilarating to head-scratching to, "Did they just ask me that?" We've waited days, weeks, and even months to hear back, only to get ghosted like a bad Tinder date—if Tinder is still a thing these days.

We've received offers that looked amazing on paper but felt wrong in the gut. We've accepted some, only to back out a few days later when clarity (or caffeine) kicked in. We've had rejections: some fair, some brutal, and some so vague you start questioning if your résumé was swapped with someone else's mid-process.

There was a time when we wondered if we should chase the so-called SHREK recruiting companies (if you think we're talking about a famous ogre, you may need to Google this): recruiting giants where prestige flows like cold brew and offers come with LinkedIn-worthy applause. The application process makes you realize something important: You shouldn't climb a ladder if you're not sure you want what's at the top.

So instead, we paused. Reflected. Built. Wrote a book. Raised four kids. Ate decadent cookies from our favorite Manhattan bakery—the kind that make you duck into a tiny street on the Upper West Side, walk out with a warm paper bag, and carry it like it's a treasure. The cookies, the size of softballs, when warmed in the oven at home, release molten chocolate that spills in ribbons, gooey enough to make the kids cheer and us forget, briefly, whatever stress put them in the oven in the first place.

We've paused, reflected, built... and we've also indulged, laughed, and savored, because delighting in life's little decadences is just as important as delighting the consumer. Through it all, we've stayed open to whatever's next, knowing it could be bigger, bolder, and more unexpected than anything we've imagined so far.

Here's What We've Learned:

- **Clarity takes time.** It's okay if you're not 100% sure what you want. Just be curious enough to explore.

- **The right job will stretch you, not just impress you.**

- **Every "no" teaches you something, especially when it's unclear why it was a no.**

- **Connections matter.** Not in a slimy way, but in a "humans helping humans" kind of way.

- **Grace and humor will carry you farther than polished answers alone.**

Therefore, to the job seekers out there: don't fear the process. Laugh at it. Learn from it. Leave room for the unexpected. Remember: Rejection isn't always redirection; sometimes, it's just noise. You're not alone. We've been there. If you ever need to vent about the application that made you upload your résumé and then type it out manually? We're listening.

By "we," I mean both of us, because while I've had my fair share of interview war stories, Yelena has her own. In fact, some of my favorites come from her early days, when she was stepping into the corporate world for the first time, armed with fresh degrees, high hopes, and (as you'll see) a few lessons learned the hard way.

Yelena's Experience With the Interview Process

Over the years, I've been part of more interviews than I can count, sometimes as the candidate, sometimes as the interviewer. I still vividly remember being fresh out of graduate school, ready (or so I thought) to dive into the corporate world.

I was wrapping up my master's degree and had decided to apply to everything and anything in the food science industry: companies all over the country, roles I understood, and roles I wasn't entirely sure about. My mindset was simple: cast the widest net possible and figure it out later. I didn't know if anyone would call me back, so why not?

I applied to more than a dozen jobs and heard back from only three. Two of them brought me through their entire interview process and ended up offering me the role.

Back in 2010—yes, I'm dating myself here—every interview was in person and not just an in-person meeting. It was a full-day production. Hiring managers, cross-functional team members, and the famous "wild card" were all part of the show—I mean, *interview*. To this day, I still question why sales always needed to interview me for an R&D role. I think they just liked getting free coffee with the new candidates.

Fresh out of graduate school, I leaned heavily on my research experience. I could talk HPLC methods, tomato chemistry, and food ingredient functionality in my sleep. However, I wasn't prepared for one particular moment.

During one of my first interviews, the interviewer asked me to explain, in detail, every component of an HPLC column. I froze. I had no idea. My brain screamed, *I'm supposed to know this*?! Instead,

I did exactly what I had been taught: I smiled and said, "That's a great question. I don't know the answer off the top of my head, but I'd be happy to follow up."

It turns out that was exactly what they wanted to hear. The question wasn't about HPLCs; it was meant to test how well you handled pressure and owned what you don't know. I ended up getting the offer. Later, I learned that asking impossible questions was kind of that person's thing. A rite of passage.

That experience taught me two things:

- Be honest about what you know, and even more honest about what you don't.

- The real test isn't whether you have every answer, but how you navigate uncertainty.

Fast forward to 2020. I was ready for a new role, and a perfect opportunity appeared. I made it through the early interview rounds, with the final on-site interview scheduled for March 17. Then, the world shut down. COVID hit. My "final interview" turned into weeks of "Let's see how things go." Weeks became months, and eventually, the company decided to hire internally. I was disappointed, but at that point, we had bigger things to manage: two kids suddenly home from school, work chaos, and the growing reality that a relocation during a pandemic probably wasn't realistic.

By that time, I had been through the hiring process of four different companies, each with its own rhythm. Post 2020, the game changed again. Interviews went virtual. What used to be an entire day on-site was now spread across multiple Zoom calls over several weeks. It wasn't uncommon for the process to stretch from application to final decision over two to three months.

Somewhere along the way, I realized that interviews are less about checking boxes and more about demonstrating who you are, how you think, how you solve problems, and whether you can lead. For me, that means being confident in my skills, sharing real examples, and sprinkling in a little humor. At the end of the day, a human being is still on the other side of the screen—at least, for now (see AI section)—and connecting with them matters just as much as answering their questions.

Yelena's Lessons Learned

- **Own what you don't know:** Interviewers care just as much about how you handle uncertainty as they do about your technical knowledge. "I don't know, but I'll find out" can be the smartest thing you say. "Here's how I will find out" is even smarter.

- **Demonstrate leadership early:** Promotions aren't earned by just checking boxes. Show initiative, problem-solving, and the ability to think beyond your immediate role, even in the interview stage.

- **Be a storyteller:** Real-world examples stick. They make you memorable long after the interview ends. Bonus points if it involves a science experiment gone wrong, a rogue tomato, or a surprisingly tense moment with an HPLC machine.

- **Sprinkle in humor:** You're more than a résumé; you're a human. A small, appropriate moment of levity can help you stand out and connect. A well-timed joke can melt the iciest interviewer. Just... maybe avoid the "So, do you come here often?" line.

- **It's a two-way street:** You're interviewing them just as much as they are interviewing you. Pay attention to how they treat people during the process, as it's often a preview of the culture.

When the Early Grind Becomes the Mid-Career Maze

José: Let's say you've already slogged through the application trenches. You've survived the entry-level ambiguity, made it past the awkward first promotions, and now you're a bit more seasoned, a decade or more into your career. It should get easier, right?

Not exactly. Landing an executive role at 41 years old is no small feat, but I've always wondered how other executives landed their positions. Was it the Ivy League degree? Was it pure grind? Was it longevity? Was it luck?

For me, it came down to three ingredients.

First, someone pointed me toward the opportunity. They had already climbed the proverbial ladder and understood how careers grow. It was, almost always, a mentor. Someone with experience, connections, and foresight to spot the opening before I could and think of me for it. The old adage still rings true: It's who you know. For better or worse, relationships remain the currency of opportunity.

Second, I had the right credentials at the right time. Back in our Employee Resource Group (ERG) days, we were told to always be ready. No one was handing out golden tickets. When the moment came, you needed the résumé, the results, and the proof that you could deliver. Passion isn't enough. You need the receipts.

Third, you must attract and hire talent, and not just any talent, but people who are, in many ways, better than you. The ones who brought skill, drive, and curiosity created the breakthroughs, the "miracles" that moved us forward. The ones who needed constant handholding didn't just slow us down; they endangered the mission. If someone has to be carried the whole way, they may not yet deserve the label "talent."

Some might call those three things "luck," but as you've seen in this book, it takes far more than luck to keep growing in a corporate system designed to exhaust even the most tireless souls.

Yet, there I was, signing on the dotted line to join a small but significant company in our industry as an executive. Did my earlier roles prepare me for leadership? Absolutely. Each one did, in its own way. Go further back in any of our careers, and you'll find that many of us started with random jobs—but we all started.

When I began my journey, my résumé featured "telemarketer" and "summer camp counselor," plus a bachelor's degree in biology with a pre-medicine minor. That's it. It also explains why I started at PepsiCo as a tech. The point is that you don't need the perfect background to start. College in the late '90s and early 2000s taught you how to learn, or at least it did for me. I left with an insatiable thirst for knowledge, so, naturally, working with sodas felt like the perfect place to try and quench it.

Here's the thing, though: Everything you're reading here is drawn from lived experience. We've seen it work, we've seen it tested, we've tested it ourselves, and we believe it's useful. Admittedly, the world changes at a rapid clip. What works in one era can be rewritten in the next. Corporate cultures evolve, societal norms shift, and the strategies that helped us succeed might look different, or even wrong, to someone reading this decades from now.

We're okay with that. In fact, we welcome it. If, in some distant future version of the working world, this memoir ends up serving as a "what not to do" guide for someone, that's still a win because it means they're thinking critically, questioning advice, and forging their own path.

Our goal isn't to give a set of rigid rules. It's to share what worked for us, hoping it sparks your own version of success, whether that means following our steps or deliberately charting a different course.

Looking back, one of the reasons I could even begin to shape my own path was because I was surrounded by people who were generous with their time, wisdom, and encouragement. I was lucky enough to have a lineup of mentors who saw my curiosity and thirst for knowledge and decided to pour into it. These scientists, managers, and leaders, all with deep food science expertise and years in labs and manufacturing facilities, didn't just answer my questions; they pushed me to ask better ones. I'll be forever grateful to them. Someday, I may write an entire book just to say thank you to all of them, but for now, let's start with one of them.

Every so often in your career, you meet someone who somehow manages to be a friend, mentor, and occasional partner-in-crime all at once. For me, one of those people is Lance. We first crossed paths during our PepsiCo days, back when corporate hallways still echoed with the clack of flip-phone text messages and meeting invites printed on paper.

Lance was (and still is) the kind of guy who can walk into a room and have people laughing within five minutes, often at my expense. He was the "handsome" sales rep from Texas, and I was the "nerdy" scientist from New York, which instantly became our running joke. Somehow, despite the miles between us, social media and group texts have kept that banter alive for years.

I've never met his kids in person, but the stories and photos suggest they're as delightful as advertised. He, on the other hand, remains my longest-running exception to the rule about screening calls, though recently, even my phone seems to be forgetting him.

Over time, Lance became part of my professional and personal circle, the kind of person you'd seek out for advice, perspective, and just enough sarcasm to keep you honest. You'll see what I mean when you hear from him directly later in the memoir.

Between Yelena and me, Yelena surely had the most convoluted path into product development. She went through the analytical function, became a liaison between operations and engineers, and even had a technical sales type of responsibility at a flavor house. There's a lesson in there to always pursue your passion, even if it takes time and the path is not clear.

Yelena: The early years at PepsiCo were a blur, not because I can't remember them, but because I was deep in the grind. I was determined to build a name for myself, especially after I began dating José, who was already a well-established product developer. I didn't want to just be seen as José's girlfriend; I wanted to be respected for my own work and seen as a smart, dependable, and capable scientist who could get things done. That meant long hours on the bench, weekends in the lab, and a relentless push to prove myself. I was never alone. We were all in the same boat: young scientists trying to earn our stripes and juggling more projects than anyone could reasonably handle. And yet, we handled them. Together.

Pepsi fostered a kind of all-consuming ecosystem. You didn't need to leave the office to live your life. There was an on-site gym, intramural sports (yes, I played basketball and volleyball, both on José's teams, which I'll talk about a bit later), and a community garden to escape into when the pressure got too high. We spent long nights mixing formulas and prepping for consumer tests, often huddled together over pizza or Chinese takeout, and playing sand volleyball in between batches. If someone had to be in on a Saturday to crank out 10 gallons of product, you could bet others would show up too, sometimes to help, sometimes because they had their own work, but mostly because we never let each other go it alone.

It was intense, exhausting, and honestly, incredible. Those four years taught me more than any textbook ever could. I built resilience and grit, learned how to navigate business and science simultaneously, and became the kind of product developer who knows every detail inside and out. I had to. It wasn't until I left Pepsi that I truly realized how foundational those years were. They shaped not just my career, but who I am.

José: The early years tend to feel less like work-life balance and more like work-life blend. In food science, stability testing runs on a calendar whether you're ready or not. Product timelines don't care about your vacation plans; they're built around pull dates, those moments when you open a stored sample to see if it's holding up to its promise.

Early on, this blend can be both exciting and bewildering. Promotions often seem to come easier, and opportunities may be tossed your way from across the industry simply because you're considered "affordable talent"—blunt, but true. The upside? You'll almost always have a job. You're also "below the radar" of upper management politics, which can be a safe, drama-free zone to learn in.

That said, the early years are also filled with unfulfilled aspirations, a search for your niche, and the temptation to measure your progress against peers. Your classmate with the lower GPA just got promoted, and you didn't? It's okay. Control what you can control. Stay curious. Resist the urge to peer over the proverbial fence; comparison only breeds jealousy or arrogance.

Another quirk of this stage: Many of us come out of college eager to give back to our communities and to the mentors who helped us. It's a noble impulse, but in these early years, you have to help yourself first before you can meaningfully help others. Midway through my career at PepsiCo, I began placing science students from my alma mater, Manhattanville College, into internships. About 10 of them went on to have successful food industry careers. They all shared the same desire to "pay it forward" right away. My advice was always the same: establish yourself first. Get your footing, build your skills, then go back and open doors for others.

The early career phase is also where professional etiquette is learned (sometimes the hard way). I'll share a personal favorite of a time when I thought I knew about formal dress code. Soon after graduating from Manhattanville, I was invited to a prestigious joint dinner between PepsiCo and my alma mater. It was held at a very upscale venue in New York City: formal, high profile, intimidating. My outfit? Dress pants and a long-sleeved button-down white shirt. Perfect for business casual... but a mile off from what was expected. I even asked the attendant at the door if I should dash out and buy a tie. He shrugged and pointed me to an expensive shop nearby. I almost went, until I remembered I didn't know how to tie a tie. So, I went in as I was. You know what? It was fine. Moments like that stick with you because they remind you how much of your early career is about learning what you don't yet know. Yes, I can now tie a tie.

Whether you're juggling an ambitious launch schedule or, years later, juggling kids' dance practice with your own deadlines, you always face a variation of the same challenge: balancing beakers and bedtime stories. The mix changes over time, but the skill, blending priorities without letting either spill over, is something you start learning in those very first years.

Did you know the Amazon River is home to over 3,000 species of fish, according to the World Wildlife Fund? I'd have called that number ridiculous until I saw it with my own eyes. I still remember wandering down to the riverbank, weaving through a maze of vendors selling fish in shapes, colors, and sizes I'd never even dreamed existed—and I say that as someone who grew up in Puerto Rico, where fried fish markets are like donut shops here in the US. *Empanadillas de pulpo,* anyone?

Now, the smell might send some people running for fresh air, but to me, the mingling aromas of river fish, sizzling oil, and crackling fires were elating. Each stall was a riot of shimmering scales, rainbow hues, and the sizzle of fillets hitting hot pans that all the grandmas used until the end of time. Despite the beautiful and powerful Amazon River being a couple of steps away, I wanted to stop at every single vendor. That's how I discovered piranha soup, which was so outrageously flavorful that I was convinced it could cure the common cold. Honestly, I didn't even mind the daily inconvenience of popping those horse-sized malaria pills for the privilege of being there, solving mysteries half a world away.

You're probably asking, "Fantastic fish tale, José, but what does this have to do with the early grind of a corporate career?" Well, during my first years on the job, I wore my business travel like a badge of honor. I felt important, chosen to trek around the globe to tackle puzzles no one else could crack. And crack them I did. Why was the grapefruit emulsion too bitter in one sample and perfectly fine in another in Brazil? Solved. Why were mysterious flecks or "flock," as we developers affectionately call them, floating in soda bottles in the Dominican Republic? Solved. Why does productivity mysteriously plummet in Puerto Rico between Thanksgiving and about eight days after Three Kings Day? Solved—but look up the words, *parranda*, *coquito*, and *fiesta* in Google Translate if you want an inside look into how I solved that one.

It was exhilarating. It was new. It was exhausting. Enthusiasm has a shelf life. Somewhere along the way, the thrill gave way to fatigue. These days, if someone asks me to travel to Chicago, I'd huff, puff, and start checking for direct flights at civilized hours, but not for myself. It's too tiring. Too disruptive. Send the newer employees, the ones still starry-eyed and eager for adventure. Let them collect the passport stamps and take the malaria pills, even though those memories and adventures will always hold a special place in my heart.

The energy you have early on in your youth, the drive that only monetary compensation can soothe, and the willingness and capacity to work late nights and early mornings are assets everyone should leverage. Work hard and play hard is a mantra we all used. I am starting to observe younger generations get anxious because they haven't found their calling, and they are, what, maybe 28 years old? I mean, come on! You have plenty of time! Take that energy and spend it like there is no tomorrow. Your old self will thank you.

Learning how urgency and importance co-exist is one of the earliest lessons in the corporate grind. You're thrown into the deep end, gasping for breath, juggling projects labeled "urgent," "important," and "mildly confusing but definitely overdue." You begin to see that prioritization isn't just a task; it's survival.

A mentor once taught me the urgent vs. important matrix, and it's worth tattooing onto your brain. If a task is important but not urgent, schedule it. If it's both, drop everything else. What if it's neither? Well, congrats, you've just found someone's pet project... or your own favorite way to procrastinate. (You know, like building the perfect color-coded spreadsheet to delay writing the report that matters most.)

Before mastering prioritization, you must know what it is you're even supposed to be doing. One of the more confusing aspects of early prioritization is understanding what your job even is. What exactly does a food scientist do? How do you explain your role to your parents, neighbors, or your kids' teachers in a way that does the job justice? Once you can answer these questions, everything else starts to make a little more sense.

Part 2: The Main Course: Inside the Industry

Chapter Five

The Chef's Special: What a Food Scientist Really Does

Sister Industries, Shared Future

O f course, it's impossible to talk about food science without acknowledging the elephant in the room, or, more accurately, the 64-ounce soda in the convenience store. Every so often, someone will say, "Food and health education in America has lots of problems." No arguments there.

When a soda that could fill a fishbowl is cheaper than a bag of grapes—unless you're at Yankee Stadium, where it costs more than your monthly electric bill—and health class teaches you more about the food pyramid than how your body actually works, it's safe to say the system could use a tune-up.

José: This chapter is all about what we food scientists do, what we are, what we aren't, and why "You must have the best job ever if you just taste gummies all day" is only partially true. Let's dive into the real day-to-day, bust some myths, and show you what happens after you prioritize your way through the urgent and the important.

Food scientists are not here to fix the whole thing in one heroic leap; that's above our pay grade and cape capacity. If we can close the gap between what people *could* be eating and what they *are* eating, one product, one habit, and one conversation at a time, then we're making progress. That's our real purpose: turning big, messy problems into smaller, solvable ones.

Purpose doesn't always arrive with a grand entrance. Sometimes it shows up quietly, in the choice to do the next right thing, again and again, until it starts to matter in ways you didn't expect.

We don't often make headlines, unless something has gone very, very wrong. Yet every day, quietly and without fanfare, we're solving problems that keep billions of people fed, nourished, and safe. We extend shelf life so food can make it from farm to table without spoiling. We fortify products to fight nutrient deficiencies that still affect millions worldwide. We make food taste better without sacrificing safety, and we expand choices for people with allergies, dietary restrictions, or limited budgets.

Yes, the food system has its flaws, but we stand firmly in the camp that food scientists, us included, do infinitely more good than harm. Every loaf of bread that stays fresh a little longer, every product that delivers nutrients to someone who needs them, every formula that makes a healthy option both affordable and delicious—those are victories.

We know it's not glamorous work. There are no cape-and-mask moments in our lab coats (though we've been tempted). We purposefully find methods to keep food safe, like pasteurization and homogenization of milk, for example, only to have consumers choose to drink unsafe raw milk because it could be healthier. Once again, scientists have failed to create a story more appealing than the charm of a healthier option supported by ancestral tradition and lore. Nevertheless, if you've ever grabbed a snack, poured a drink, or cooked dinner without worrying whether it was safe, fresh, and nourishing, then you've already been on the receiving end of what we do. We think that's worth celebrating.

The food and beverage industry and the vitamin, minerals, and supplements (VMS) industry have long been considered cousins at best: similar science, different family reunions. Recently, though, they've started blending in ways that reflect a deeper shift. Consumers aren't just looking for taste or nutrition anymore. They want both, and they want it with purpose.

I was fortunate to work for a company that straddled both worlds. I led a lab that supported VMS and food and beverage innovation simultaneously. It was intense, but invaluable. What stood out most was how the scientific mindset, and even the air in the room, had to shift when switching lanes.

On the food and beverage side, the space hummed with mixers churning colorful liquids in stainless steel beakers, while refractometers and hydrometers measured Brix, the number of solids in the liquid, to nail the sugar content. Fishy-smelling gummies, ice cream streaked with beta carotene, and high-speed mixers filled the lab with a smorgasbord of sights, sounds, and smells. The work was as much about flavor and mouthfeel as it was about chemistry.

Step into the VMS lab, and the sensory world narrows. Instead of bright aromas, the muted scent of raw powders greets you. The machinery speaks in different terms: friability testers tumbling tablets to check for breakage, hardness testers applying steady pressure until they snap, and V-blenders slowly folding powders to ensure uniformity, which is important to ensure that every capsule or pill has the same amount of nutrients in it. Here, success isn't judged by "refreshing" or "creamy"; it's determined by stability, potency, and whether a capsule can survive a cross-country shipment intact.

Going from a beverage formulation to an oral solid dosage form isn't just a change in format; it's a full mental recalibration. The possibilities of leveraging each other's strengths can lead to extraordinary results, like beverage-based supplements that taste good (a miracle, frankly).

In the beverage world, you're checking brix, acidity, and whether your formula will hold together long enough to survive a summer on a store shelf. In supplements, you're running friability and hardness tests, making sure a tablet survives the trip from factory to customer. Then there's zeta potential, the lab's way of asking, "Are all the tiny particles in here playing nice?" A high zeta potential means they're keeping their distance, staying in smooth suspension. A low one? Think of a bad wedding where all the guests start clumping together in a corner until the whole thing falls apart.

One area where the crossover is still surprisingly limited is sensory science. In food and beverage, taste, aroma, and texture are make-or-break. In VMS, they've historically been an afterthought, often sacrificed in the name of function. Yet the sensory challenges in VMS are tenfold: bitterness, metallic off-notes, dissolving speeds, pill coatings that stick to your tongue... the list goes on. From what I've seen, the talent pipeline hasn't caught up.

If you're a scientist thinking of jumping from one side to the other, know this: The science will travel with you, but the priorities will change. In food and beverage, the consumer's first sip decides the product's fate; in VMS, it's often the stability data six months later that makes or breaks a launch. Both worlds reward precision, creativity, and problem-solving, but they speak different dialects. Learning both is like being bilingual in science, and it's a skillset few have but many will need in the future.

So, dear reader, if you know someone, maybe a student exploring colleges or a curious kid who lights up at the science of flavor, consider guiding them toward food science and sensory science. It's a field where two industries meet, where innovation still has plenty of room to grow, and where the next generation can make something not just functional, but delightful.

Early Retrospective

Yelena: My parents thought I worked in a factory. For years. No matter how many times I explained beverage systems or appliance and beverage technology, they still pictured me in a white coat on a production line, hairnet and all, taste-testing cherry soda like a glamorous lab rat.

José: Mine were slightly more forgiving. When I turned down a research position at Mount Sinai to work for PepsiCo for the exact same pay, $35,000 a year, I could feel my mom glare at me through the phone, like I had just chosen to backpack barefoot across the

Bronx. She didn't know what food science was, but she did know she wanted me to wear a white coat and cure something.

Yelena: Here's the thing: We do wear lab coats. We just swap out stethoscopes for refractometers, and instead of saving lives (at least directly), we make sure your slushy doesn't taste like watered-down cough syrup and that your juice doesn't separate into something that looks like lava lamp goo.

José: My mom's reaction when I told her I'd chosen PepsiCo over med school was one of unamused encouragement, equal parts support, and silent questioning. Her son, the future doctor, is now working on... soda? But my parents had always taught me to "be the best at whatever you do, no matter what it is," and while they didn't understand food science, they never withheld their belief in me. That belief—quiet, steady, unglamorous—helped me fuel what I can only describe as a fire in my belly.

But what was I chasing? At first, I thought it was science. The way acids and bases collided into something refreshing (if added in the correct order, of course). The zeta potential of emulsions. The seductive smells of a sugar-scented lab. It all thrilled me, but now I realize that wasn't the whole picture.

Looking back, I was drawn not just to chemistry, but to what chemistry built. I didn't have the language for it then, but I was fascinated by how a simple formula could scale into a billion-dollar product. Somewhere between pipettes and profits, there was a kind of magic. One I longed to master.

Product development, I've come to believe, is a form of creation that borders on parenting. We joke in the industry that our launches are like our babies. "Look at her, all grown up on the shelf!" we say. And while the metaphor makes us laugh, it also holds some truth. You pour your ideas, your energy, and your late nights into something that didn't exist before, and then you let it go, hoping it thrives.

The irony is that in those early years, I didn't see myself as purposeful. I was following curiosity, instinct, and opportunity. But swim against the current long enough, and you start to realize there's a direction forming beneath the drift. What once felt aimless begins to look, in hindsight, like a map.

And while I never lost my doubt (if anything, it matured), I began walking into work with more intention. I wasn't just solving problems. I was learning to build, to lead, and to connect dots that weren't obvious. The lab became a training ground, not just for formulation, but for formation.

During those first seven years at PepsiCo, walking into work often felt like stepping onto a stage: one part scientist, one part showman, all parts problem solver. The anticipation was electric. Whether it was crafting a formula that danced on the tongue, hitting a problem-solving flow state Sherlock Holmes would envy, or tossing the perfect lab bench quip, these weren't just workdays. They were symphonies of creativity, camaraderie, and caffeinated alchemy. (I was working on Diet Code Red at the time.)

I didn't know it then, but I was starting to build more than products; I was building a philosophy. And slowly, that philosophy began to shape me back. Purpose didn't arrive all at once. It emerged through repetition, through knowing the machines like muscle memory, through realizing that I didn't just enjoy the work—I understood it, as if it were written in my native tongue.

At home, I can barely unscrew a lightbulb without Googling which way to twist it, but in the lab? I could disassemble and reassemble a high-pressure homogenizer with my eyes closed, thanks to mentors like Peter Given, a respected industry mentor who shared generously with eager minds like mine. Product development isn't a job to me; it's second nature. Whether refining a drink or analyzing supplement data, I understand the process instinctively.

To others, though, it's a riddle wrapped in a lab coat. Misconceptions about what we do are everywhere. Some discover the world of food science and attempt to force their way into its folds, enchanted by its promise. But like any craft worth mastering, it resists the inauthentic. If you don't love what you do, don't do it. That may sound trite, but aligning your profession with your inner rhythm, that sacred harmony between ability and joy, isn't just the secret to a great career; it's the essence of a meaningful life.

Here's something many people don't realize: Food scientists don't write recipes. We create formulas. Culinary chefs craft recipes, edible poems composed with intuition and flair. But we, their molecular cousins, are formulators. One more time, for those in the back: We are formulators.

Now, if you've noticed us dropping culinary terms throughout this book, yes, we know what we're doing. It's part of the fun. We call a chapter "The First Bite" not because we're teaching you to make bruschetta, but because we want you thinking about the opening bites of a meal, those first tastes that spark anticipation. We want you to feel like you're in the kitchen, even though we're standing next to a high-shear mixer or microfluidizer.

That's the twist: Food science isn't culinary... but we'll flirt with that confusion all day long. Because once you understand the difference, you'll also understand the connection. Both worlds speak the language of flavor, texture, and transformation. Only, in the kitchen, the "salt" might be a mineral premix, and the "bake time" might be a 72-hour stability test at 40°C. Bon appétit!

The chef's job is to build what's known as a Gold Standard, the platonic ideal of a product. Imagine the perfect beverage—vibrant, flavorful, unforgettable—except it spoils in one day, and it can't be mass-produced. That's where we come in.

Edgar Allan Poe argued that great work begins with a single intended effect and is constructed backward to preserve it. That's our blueprint too. The Gold Standard is the "effect"; product development is the composition. Our mission is to translate a fleeting, chef-driven moment into a formulation that is safe, legal, stable, efficacious, and scalable, without losing the original spell.

Developing safe products is paramount. Developing stable products is why we're hired. You learn which levers to push or pull, whether it's for pH, water activity, emulsifiers, or sweetener systems, to keep that matrix coherent for its entire shelf life. In other words: Keep the evocative ballad's emotion intact, even when you must change the meter.

When you hear "keep it stable," you might think about avoiding mood swings. In product development, it means adjusting acidity, sweetness, or viscosity until a beverage stays perfect from Day One to its printed "best by" date. No separation, no off-notes, no surprises. You also need to know as much about the product as you do about the package. Great taste is what separates the stars from the superstars. If you've heard of high-pressure pasteurization (HPP), you know it uses high pressure to kill pathogens and extend shelf life without altering food taste or quality. But making something taste amazing? That's a different kind of pressure.

We replicate both the flavor and the spirit of the original, creating something that can survive the supply chain, please regulators, and still delight the consumer. It may not be a mirror image, but it's the closest thing to perfection that can thrive on a store shelf. If we've done our jobs right, it still sings, just in a slightly different key.

Yelena: This is where José and I differ in our approach. He's the engine behind the formulation; I'm the tuning fork. I live for subtleties: the scent that disappears too quickly, the note on the back of the tongue that feels just slightly... off. I obsess over how texture, aroma, and balance come together to satisfy, surprise, and most importantly, transform.

To me, the Gold Standard isn't just a starting point; it's a sacred experience, a memory made tangible. I love to cook, and cooking is where I train my senses. I bring the same mindset into product development. My job is to make sure that when someone opens a bottle or chews a gummy, they feel something: comfort, curiosity, delight. The tiniest tweak in acid level, the faintest hint of vanilla, the decision to round off an edge rather than sharpen it, the addition of rum flavor to a pineapple because the dark caramel notes harmonize with the sweet tropical notes—that's the magic. That's where I live.

So, yes, we are formulators. We're also dream interpreters, taking the vision of an ideal product and coaxing it into reality, shelf-stable and stunning. When we get it just right, the science fades quietly into the background, leaving the consumer with only one thought: *Wow, that's good.*

José: Back when we worked with PepsiCo, the Chief Science Officer would occasionally speak to the local IFT Chapter. At the time, the word *natural* was starting to attract both consumer fascination and industry scrutiny. It was a buzzword in desperate need of a definition, and depending on who you asked, it still is today.

I remember the speaker addressing the crowd and saying, in words I'll paraphrase: "Snake venom is natural. That doesn't make it healthy." Then, pacing across the stage, the speaker drove home another point: "Science doesn't magically change when you cross an imaginary border between countries." He stepped one way and said, "Legal here," then stepped back and said, "Illegal here."

Those words have stayed with me. They are a reminder that science has to help itself if it's going to regain and keep the public's trust.

I know, even as I write this, some readers are rolling their eyes, thinking, *Yeah, right. Good luck with that.* It's easy to be cynical, especially in an age when TikTok and other social platforms have amplified a wave of science denial. Many of us in the field feel under attack. Funding cuts and restrictions on academia have eroded morale. If you want a pulse check on how scientists are feeling these days, just scroll through LinkedIn. The frustration is palpable.

And yet there's another truth we must name. As Neil deGrasse Tyson often points out, science can become a victim of its own success. Vaccines work so well that the diseases they prevent fade from memory; when the fear disappears, skepticism rushes in to fill the vacuum. If you've never seen a polio ward—and I suspect none of us have—"natural" starts to sound safer than it is. That's not proof against science; it's proof that science must tell its story as clearly as it does its math.

This is why Yelena and I are still hopeful. Historian and best-selling author Yuval Harari writes that science has a self-correcting mechanism: Over time, bad ideas lose to better ones, and evidence wins the rematch. The peer-review process is flawed, human, and slow, and yet it is still one of humanity's most powerful tools for truth. We trust that science, and scientists, will find their way back into public favor, not by shouting louder, but by making the evidence legible, memorable, and human.

Speaking of finding our way back to our senses...

Food science isn't just about beakers, formulas, and lab coats. It's an ecosystem of specialties that work together to turn ideas into safe, successful products. While most people are familiar with sensory science, regulatory affairs, and quality assurance, there's a whole menu of lesser-known yet equally vital paths, like becoming a thermal process authority (yes, that's a real title) or managing complex product launches from within R&D, engineering, product development, and quality assurance functions.

Let's start with our friends in sensory science.

You've probably heard the phrase, "Taste is King." It's one of the simplest truths in product development, especially when you're working on sensorial formats like beverages, gummies, and bars. Time and again, consumers vote with their wallets, and the message is clear: If it doesn't taste good, it won't sell, no matter how great the nutrition panel looks.

There's a long-standing belief that for supplements to be effective, they need to taste bad, as if bitterness is a proxy for potency. While there may be a kernel of truth in that perception, there are plenty of wildly successful brands with minimal active ingredients that thrive on one thing alone: great taste. There's a sweet lesson in there somewhere.

Sensory science plays a critical role in food and beverages, but it's still underutilized in the dietary supplement world. This needs to change. Understanding how to properly conduct sensory evaluations is crucial for success, especially as consumer expectations evolve.

These types of statements invite regulatory scrutiny and should be backed by robust, independent, and well-designed sensory data.

I also led the Regulatory function in my last role as Vice President of R&R, which, despite what it sounds like, did not stand for rest and relaxation (though wouldn't that have been nice?). In this case, R&R meant R&D and Regulatory, and I quickly learned that if R&D required translation for the business, Regulatory needed full-blown subtitles.

Striking the right balance on the regulatory pendulum has been one of the toughest and most nuanced challenges of my career. For example, the tighter the regulatory grip, the more protected a brand can be. Swing too far and you risk slowing innovation to a crawl, nay, a standstill. On the flip side, take too loose an approach and you might find yourself with a label claim that keeps the legal team up at night.

My guiding principle became this: As the risk and the size of the prize go up, so too should regulatory rigor. If a product hasn't launched yet, sits in the early R&D stage, or just launched but under a smaller brand name, the risk is lower, so long as it's safe, legal, and stable. The business can and should explore a bit more freely as it goes. Iteration is the backbone of innovation, but once a product gains traction or becomes tied to a flagship brand, the moat needs to rise. You're not just protecting a Stock Keeping Unit (SKU); you're protecting equity, reputation, and shareholder confidence.

Now, I can already hear my regulatory friends saying, "Easy there, cowboy," and I get it. This isn't carte blanche to throw caution to the wind, but business leaders often need to be reminded that Regulatory isn't the "no" department. With strong partnerships and sound scientific rationale, they can be powerful enablers.

Pro tip: To find your company's regulatory "center of the pendulum," look at your R&D team's capability to craft logical, evidence-backed arguments. The stronger the team and the closer their connections to functions like quality, the more flexibility you may have, but always within the bounds of responsible science and good judgment.

Picture the pendulum, specifically a pendulum swinging between two extremes. On one end, it's tight and stifles innovation: Every claim requires three peer-reviewed studies and a papal blessing. On the other end, it's too loose and launches with bold claims like, *Makes you immortal!*—followed by a recall notice and a PR disaster. So, where's the sweet spot? At the intersection of science and strategy, of course. Risk is managed. Innovation flows. The brand grows, and you get invited back to the next team meeting with open arms and a new reputation as an innovation leader. Get the pendulum tilted too far to either side, and you may be treated like a bad case of the flu. No one wants the bug... except maybe microbiologists. That leads us to our next thought.

Mind Your Bugs: The Unsung Hero of Food Science

Microbiology. The word alone reeks of boredom, like something you'd only study if your dream job was swabbing surfaces in a lab coat that hasn't seen sunlight since Y2K.

Here's the truth: Microbiology is the secret backbone of product development. It's what keeps your snack shelf-stable, your beverage safe, and your lunch from turning into a biohazard.

As developers and formulators, we don't just make food taste good; we make sure it doesn't fight back. We're trained to build microbial hurdles that keep spoilage at bay and pathogens from turning a product launch into a public health crisis. We must know the difference between pathogenic *E. coli* (the one that sends you to the ER) and non-pathogenic *E. coli* (the lab rat of microbes). Are you kidding me?

Let's not forget our old frenemy, *Clostridium botulinum*, which is deadly in a can, but somehow totally acceptable when injected into your face for wrinkle reduction. (Food scientists: 1, dermatologists: also 1?) The duality is wild!

For me, the glamour of food safety took shape in places like Denton, Texas, where I flew out (more times than I can count) to ensure our aseptically processed, packaged beverages were microbiologically sound. Sure, I racked up miles and added to that badge of honor I mentioned earlier, but I also earned the right to say: "Those drinks won't kill you. You're welcome."

If you were wondering, yes, I did get to eat at that rotating restaurant in Dallas. The one with the chef-dream steaks and skyline views that drift past the windows at an unhurried pace, while the deadlines back home are the only things spinning fast. Worth it. These days, though, I'm retired from volunteer travel duty. That baton is ready to be passed.

Pro Tips for Aspiring Food Scientists

First, if you're new to formulations, mind your bugs. They may be microscopic, but they will humble your million-dollar idea faster than a bad Yelp review.

Second, if you're working with anything above a pH of 4.6, I've got two words for you: Process Authority. Get one. Seriously. Your product, and your peace of mind, depend on it.

One of the lesser-known parts of being a food scientist is working with standards of identity. Think of these as the legal definitions of what certain foods must be if they're going to carry a familiar name on the package. Ice cream has one. So do milk, chocolate, jams, and jellies. If a product doesn't meet the standard, you can't legally call it by that name, no matter how close it tastes.

On the surface, it sounds bureaucratic, but these standards exist to protect consumers. When you buy "ice cream," you shouldn't end up with a frozen dessert that has never seen a drop of cream. When you buy "chocolate," you expect actual cocoa solids, not just brown-colored sugar. These guardrails make sure words mean something in the grocery aisle.

Both of us have had to wrestle with standards of identity. Sometimes it's a headache because your brilliant new formulation might not meet the standard, even though it's safe, nutritious, and delicious. More often, it's a chance to innovate within the rules, to find creative solutions that respect tradition while moving the industry forward.

That's the hidden lesson: Food science is not about cutting corners; it's about building trust. Behind every carton of milk, every bar of chocolate, and every jar of jam, an invisible safety net of science and regulation ensures that what you see is what you get.

One of the most persistent and frustrating myths we've encountered is that dietary supplements aren't regulated. We've heard this from consumers, journalists, and, most disappointingly, from highly respected scientists in adjacent fields who should know better. The truth is: Supplements are heavily regulated. They're just not regulated in the same way as pharmaceuticals, and that's where the confusion begins.

Let's break it down.

Dietary supplements fall under the jurisdiction of multiple government agencies, each playing a distinct and critical role. The FDA (Food and Drug Administration) is responsible for enforcing 21 CFR Part 111, which sets requirements for how supplements are manufactured, packaged, labeled, and held. These standards are rigorous— think GMPs (Good Manufacturing Practices)—and involve mandatory documentation, identity testing of ingredients, and the obligation to investigate and report adverse events.

Then there's the FTC (Federal Trade Commission), which works alongside the FDA to regulate how supplements are advertised and marketed to consumers. If you make a claim—say, "supports memory" or "boosts immunity"—you'd better have competent and reliable scientific evidence to back it up. Claims without substantiation can lead to warnings, fines, or lawsuits. Ask anyone who's been on the receiving end of an FTC letter; it's not a "light-touch" industry.

Even the industry itself has watchdogs. The National Advertising Division (NAD), part of the BBB National Programs, actively investigates and challenges misleading claims. When companies get it wrong, the NAD doesn't just wag a finger; it makes public recommendations that can result in significant brand damage.

All of this is grounded in the Dietary Supplement Health and Education Act of 1994 (DSHEA), which legally defines what a supplement is and sets the framework for how it's regulated. It's not a free-for-all. Supplements are a distinct category of regulated products with their own rules, responsibilities, and enforcement mechanisms.

So, where does the confusion come from?

It often stems from comparing supplements to drugs, which require pre-market approval and extensive clinical trials. Supplements, by contrast, don't require FDA pre-approval, but that doesn't mean they're unregulated. They're post-market regulated, meaning companies are responsible for compliance from the start, and the FDA steps in if there's noncompliance or risk to public health. That's not a loophole; it's just a different system, designed for a different category of products.

Here's something rarely discussed: who gets studied, and why.

In the US, supplements can't be marketed to diagnose, cure, mitigate, treat, or prevent disease. Those are drug claims. Design a clinical trial in a disease state, or to treat/prevent a disease, and you've changed the intended use. Now you're in drug territory.

Quick Primer:

- Design a human study to treat/prevent disease → you're in IND territory (Investigational New Drug; drug rules).

- Plan to market a supplement with a new dietary ingredient

→ You file an NDIN (New Dietary Ingredient Notification) less than 75 days before launch (supplement rules).

Different tools, different jobs.

In practice, that's why most supplement trials are run in healthy adults with structure/function outcomes (e.g., "supports memory," "helps maintain healthy blood pressure already within the normal range") rather than disease endpoints. Different goals, different guardrails.

Are there bad actors? Of course, just as in pharma, finance, and every other sector. Bad actors don't negate regulation; they underscore the need to enforce it.

If you're a consumer, be skeptical, but be informed. Look for transparency. Look for brands that test their products, explain their ingredients, and take their responsibilities seriously.

Rest assured: Dietary supplements are not the regulatory Wild West.

Speaking of regulation… let's talk about a product that lives in the hazier corners of oversight: Cannabidiol (CBD). Here's a chill story about how food science finds its way into the unlikeliest places.

The Great CBD Expedition: Food Science Goes Grassroots

(With commentary from the Home Office a.k.a. Yelena)

José: Back when CBD was being infused into everything from sparkling water to pet chews, I was part of a cross-functional dream team sent to explore the landscape. We were scientists, marketers, and business folks on a mission to understand this booming trend from the ground up, literally. The formats were squarely in our food science wheelhouse: gummies, tinctures, beverages, and even the occasional chewable tablet with questionable mouthfeel.

The team clicked instantly (think Ocean's Eleven), but instead of stealing a casino vault, we were extracting cannabinoids and evaluating suppliers. We traveled to places like Montana and Colorado, toured pristine hemp farms, and grilled suppliers with the intensity of a peer-reviewed journal club. In under a month, we could tell you what mattered in the CBD space, who was doing it right, and which companies were just blowing (hemp) smoke.

How about back home? That's where the real hero was.

Yelena: Let's be clear. While José was off in the mountains talking about the differences between CBD, Cannabinol (CBN), and Cannabigerol (CBG), I was in the trenches. Daily family calls, dinner prep, school pick-ups, and refereeing debates about screen time limits ("But he got an extra six minutes yesterday!"), all while holding down my own full-time food science job. At one point, I think I said "no" 47 times in one afternoon. Resiliency score: off the charts.

José: We worked hard during the day and laughed even harder after hours, solidifying friendships that still hold strong. That mix of functional expertise, curiosity, and mutual respect allowed us to gather insights quickly and present clear, confident recommendations. Unfortunately, the pandemic hit shortly after, and with it, the momentum behind the CBD market slowed to a crawl. (Timing really is everything.)

Still, that experience reminded me of the best parts of working in cross-functional teams, and the power of food science to translate a wellness trend into real, consumer-ready products. We didn't just chase cannabinoids; we built a team that could've launched anything for our customers. As Yelena says: We did all that, while she kept our actual home team running.

Misconception: "Family Has No Place at Work"

Sometimes, "we're family" is used to blur boundaries and justify overwork. That criticism is fair. But we've lived another reality: In the hardest moments, coworkers can become something more than coworkers.

On September 11, 2001, I was in a training room in Somers, New York, eyes glued to a TV on a rolling cart. Then it hit—the second impact, live. We were dismissed to be with our families. In the days after, colleagues who barely knew each other coordinated rides, brought meals, sat in silence, and shouldered workloads for those who were grieving. No one weaponized the word "family." People simply acted like one.

I've watched that instinct surface during illnesses, house fires, and private crises. The right question isn't "Should we call this a family?" It's "Do we show up like one?" Boundaries matter. So does care. Healthy cultures hold both truths. If "family" feels loaded, use "community." The point stands: Labels don't build trust; behaviors do.

As AI and other forces reshape the workplace yet again, we'd be wise to remember that while tools and processes will change, the human bonds we build, the ones forged in the moments that matter most, are still what make work worth doing.

Let's end with what a food scientist's work actually is, in one breath:

We take a chef's Gold Standard and keep its effect intact at scale.

We're not just in the grocery aisle. We're in distilleries and fragrance labs, pet food plants and skincare lines; we upcycle side streams, design barriers so oxygen and light don't steal flavor, and keep fermented things friendly in the wild. If it must be safe, stable, and irresistible, we're probably in the room.

How that happens varies by product:

- Aseptic processing sterilizes the product and package before sealing. Long shelf life, ambient stability.

- ESL (extended shelf life) keeps the product cold from fill to fridge. Shorter shelf life, fresher taste.

- HPP (high-pressure pasteurization) uses pressure (not heat) to inactivate pathogens. Great for juices and "fresh" profiles.

That's the job: pick the right lever so safety, quality, cost, and taste survive time, transport, and temperature.

So yes, we are formulators. We're also dream interpreters, taking the vision of a product and bringing it into reality, shelf-stable and stunning. When we get it just right, the science fades into the background, and the sip remains. The consumer is left with only one thought: *Wow, that's good.*

Of course, work doesn't always involve those defining moments. Most days, it's far less dramatic, a mix of routine, problem-solving, and the occasional coffee-fueled miracle. This brings me to one of my favorite contrasts: the difference between what people think a food scientist does and what we actually do. Here's a quick visual comparison, just to set the record straight:

What People Think a Food Scientist Does vs. Reality

Perspective	What They Think I Do	What I Actually Do
My Mom	Wear a lab coat in a factory, taste-testing orange soda	Translate sensory data and claim language into PowerPoints at midnight
My Friends	Eat snacks all day and invent wild new flavors	Fight spreadsheets, lead stability meetings, and reformat Excel cells
The Internet	Genetically modify food to make strawberries the size of watermelons	Reformulate to avoid titanium dioxide and still meet the cost of goods
The Marketing Team	Spend my days with other R&D ivory-tower scientists, magically turning "whispers of botanical complexity" into a drink overnight.	Spend my days more like Dr. Frankenstein, but with better lighting, fewer lightning bolts, and a knack for bringing marketers' "babies" to life in a way people will actually want to consume
The Supply Chain Team	Delay launches for fun	Rework formulas five times to match an ingredient spec no one can source
My Kids	Make candy	Technically true, on a good day
Us (José & Yelena)	Have the coolest job in the world	Have the coolest job in the world (but with a lot of email chains)

Perception vs. Reality: what people think a food scientist does

Next up: how an idea becomes inventory—*from recipe to plate.*

Chapter Six

From Recipe to Plate: The Art and Science of Product Development

José: Here's my unpopular opinion: Product developers are the artists of food science. Not "fine art on a blank canvas" artists, but constraint artists. Our pigments are acids, bases, emulsifiers, and barriers; our brushes are mixers and microfluidizers; our gallery is a bottling line running 926 cans a minute. We're not the deepest scientists in the room, and that's the point. We live at the crossroads of science, art, regulation, and consumer psychology.

Yelena: I roll my eyes when he says "artists," but he's not wrong. Art still needs a frame. That frame is *feasibility*. Before we pick a palette, we draw the borders of the canvas: what claims we can legally make, which processing route keeps the effect intact, which ingredients actually exist at volume, and whether the line and the people can run it without breaking.

José: I used to think the work began at the bench. It doesn't. It begins with questions.

Yelena: The right ones.

The Feasibility Stage: Where Vision Meets Reality

The most underestimated part of our job? The feasibility stage. Before a single droplet is mixed, we take ownership of feasibility like a chef owns a kitchen. It's where strategy meets science. A tight Design of Experiment (DOE) here can shave weeks off development timelines, saving costs, preserving momentum, and sometimes, keeping projects alive.

José: Once, in just eight weeks, I helped our team replace grapefruit oils within the flavor system after one was suddenly flagged due to emerging crop health concerns. That project not only allowed us to keep the product on the market, but we actually improved its taste scores and purchase intent. We succeeded because we were calibrated. We used the high-performing team formulation mentioned in Chapter 2: positive management, a curiosity-filled mindset, a touch of humor, and absolutely no ego.

I played my role by navigating emulsifiers, adopting novel texturizers suggested by experts on the team to match pulp mouthfeel, and balancing micro-stability and sensory delight. That's the mastery needed: emulsions, flavors, microbial safety, and package-product interaction. It's all part of the job.

Yelena: What José isn't saying is that for every project like that grapefruit replacement, there are ten others that test your patience and precision. I've worked on slushies, still beverages, and formats that don't behave like water, and formats that want to separate, curdle, or go cloudy. Each format demands different bitter masking systems, pH balancing acts, and taste-masking strategies. In one case, I had to turn what tasted like a botanical garden soaked in dish soap into a fruit-forward, kid-friendly drink. It took 17 iterations and a willingness to keep going when most would stop.

From Paper Formulas to MVPs

We start with a paper formula, often paired with a gold standard. Our job is to match that ideal experience as closely as possible, except ours must be legal, shelf-stable, and manufacturable. That's where expertise kicks in. Anyone can mix powders and hope for the best. We architect systems.

The bench stage doesn't involve random tinkering. It's iterative, hypothesis-driven experimentation using DOE frameworks to guide us toward ideal outcomes. We test stability, sensory attributes, and feasibility, all while juggling internal timelines and external expectations. We taste, adjust, and taste again.

Yelena: In many of my projects, that gold standard is more of a moving target than a static ideal. You chase it knowing you may never hit it exactly, but the closer you get without compromising stability, the stronger your final formulation. It always amazes me how something as subtle as changing the flavor supplier or the timing of ingredient addition can change everything. You learn the formula—how it should feel, look, move, and even smell—both in the lab and in the tanks of the pilot plant.

The Pilot: The Truth Teller

Next comes the pilot, a bridge between the benchtop and the factory floor. It's not just a scale-up; it's a crucible. Did your formula hold up? Are your specs tight enough? Is this variant production-friendly? Pilots don't fail; they reveal. A "failed" pilot simply means you've uncovered a weakness before it costs the company millions. That's a win.

When pilots succeed, we may go straight to consumer studies, if we're developing an experiential format like beverages, or a clinical study if we're developing an oral solid dosage form like a tablet. In consumer studies, developers obtain Just About Right (JAR) scores, purchase intent, and overall acceptability. In clinical studies, we obtain the necessary data to substantiate claims. These aren't just marketing tools; they're data that guide final decisions. They help us tell a techno-commercial story: a narrative supported by science and tailored for retailers and consumers alike.

The Final Mile: Regulatory, Stability, and Launch

A product isn't real until it survives the final gauntlet: a full-scale manufacturing run. At this stage, we execute another round of stability testing, this time in the final package, final formula, and final process parameters. Only then can we validate shelf-life claims. It's a scientific, regulatory, and commercial requirement. The package, ingredients, and process must all match. No shortcuts.

Why It Matters

This process, meticulous and invisible to most, requires a rare blend of technical mastery, project management, stakeholder engagement, and consumer empathy. It's not just what we do. It's who we are. Whether we're reformulating a juice beverage to stay in market or pioneering a new gummy supplement format, we do it with precision, speed, and conviction.

We got good at this, not because we said so, but because we earned it every time we beat the timeline, cracked the flavor code, or kept a product alive through sheer will and deep expertise. Our teams knew the rhythm of development like a jazz ensemble. Everyone had a part, and we played ours with the kind of finesse that only comes from living and breathing formulation.

Yelena: I'd add this: The real pros never stop learning. It's not about locking into a single style; it's about growing with your palette (and your palate), the data, the consumer, and the challenge in front of you. Whether it's a slushie, a hydration shot, or a protein beverage, there's no autopilot. Every launch is an orchestra. The better tuned you are, the better the music. But an orchestra is only as good as its instruments and the people who make them. In our world, those makers are suppliers and contract manufacturers.

The Secret Ingredient We Almost Missed: Suppliers

José: For the longest time, I treated suppliers like vending machines. You put in a purchase order; you get an ingredient. Transaction complete. Efficient? Sure. Smart? Not even close.

Yelena: We learned quickly that suppliers are so much more than line items on a spreadsheet. They're underutilized partners, sometimes the very difference between a project fizzling out and a launch that turns heads at Expo West.

José: Think about it. They've got global insights we don't always see from inside the corporate walls. They've got specialized equipment that we'll never buy for just one project, but they've already mastered. They've got scientists who live and breathe their ingredients the way sommeliers obsess over wine. And when you bring them into the process early, they don't just add value—they multiply it.

Yelena: Now, suppliers aren't magicians. They don't exist to swoop in at the last second and rescue a doomed formula (though more than once, they've done exactly that for us). They work best when treated like collaborators. Respect their intellectual property (IP). Be transparent about your goals. Invite them into ideation, not just execution. Too often in consumer packaged goods (CPG), suppliers are treated like the help when they're actually the hidden levers of innovation. We've seen it in both the food and beverage and the supplement industries: beverage labs leaning on flavors and gum or starch blending technologies; supplements leaning on bioactive suppliers to provide potency and homogeneity expertise. In both worlds, the magic happens when you blur the lines and let supplier expertise sharpen your own.

Our appreciation for suppliers didn't happen overnight. For me, it took joining the supplier side to fully understand just how much value they bring and how often we overlook it. When you've spent your whole career on the CPG side, it's easy to think of suppliers as external and transactional, or simply a means to an end. However, when I stepped into a B2B (business-to-business) role myself, everything changed. I wasn't just the one requesting samples anymore. I was the one racing to deliver them under impossible timelines and asking a million questions, not to be annoying, but because getting the ingredient right truly depended on it.

Seeing Both Sides: From CPG to B2B and Back Again

Working across both the CPG world—the companies creating the products you see on shelves—and the B2B, or supplier side of the food industry, has shaped me into a more well-rounded product developer and leader than I ever expected. It's given me a unique perspective on what it takes to bring products to life and to sustain partnerships at every step.

Like most developers, I learned how to interact with suppliers from watching my peers. When I first started at Pepsi, I was introduced to sales reps and suppliers by tagging along on meetings and shadowing conversations. The rule seemed simple: keep it brief, ask for what you need, move on quickly, and get back to the "real" work. After all, there were so many suppliers to manage, and the idea was to keep those interactions efficient. That mindset followed me to my next CPG role as well: clear requests, minimal back-and-forth, done.

Then I joined IFF, stepping into the supplier side for the first time, and everything I thought I knew changed. Suddenly, I was the one fielding requests from CPG developers, sometimes with timelines as short as a week. There was no room for guesswork, and no time for mistakes. And that's when it clicked: those lengthy forms and those endless questions suppliers ask? They weren't just red tape. They were the foundation for creating ingredients that work.

Take flavors, for example. When I was on the CPG side, if I needed a mango flavor for a new beverage, I'd just request "mango" and move on. But at IFF, I learned that "mango" isn't one-size-fits-all. Flavors are complex. Suppliers need to know everything: the base formulation, the processing method, and whether the product is heat-treated, shelf-stable, or refrigerated. All those factors affect how the flavor performs. A mango that works beautifully in a cold, fresh juice might fall completely flat in a heat-treated beverage with a 12-month shelf life.

Even more fascinating? Mango isn't just mango. Some mango profiles are a hit in Asia, others in South America, and entirely different ones resonate with consumers in North America. Preferences for sulphury, cooked, tropical, and candy-like all depend on the consumer target and the product vision. The more suppliers understand, the better they can craft an ingredient that hits exactly the right note.

That experience completely reframed how I see supplier relationships. It's not about sending an ingredient request and moving on; it's about collaboration. The best products are born when CPG developers and suppliers work together as true partners, understanding not just the "what" but the "why" behind each decision.

By the time I returned to the CPG world at Welch's, I became an ambassador for that partnership mindset. I approached suppliers differently, involving them early, sharing more context, and building solutions together. The goal wasn't just developing a formulation, but creating a product consumers would love from the first sip to the very last, no matter when they pulled it off the shelf.

That shift, from transactional to collaborative, has shaped the way I lead, the way I develop products, and the way I build teams. I now understand that when both sides win, the consumer wins, and that's the ultimate goal.

Winning over the consumer isn't the only challenge. Inside the walls of a company—any company—you're also navigating titles, turf, and the sometimes-unspoken rules of the organizational chart. Unlike technical specifications or formulations, no one hands you a guidebook for that part. So, before you grab your lab coat or sharpen your PowerPoint skills, let's take a quick detour into the wild terrain of corporate culture.

Chapter Seven

The Kitchen Maze: Navigating Titles, Turf, and Townhalls

José: The following chapter will be quick but crucial if you wish to navigate the pernicious currents of corporate life, and trust us, they're as real and as potent as any acid-base reaction we run in the lab. Every company has a culture and set of working practices that aren't captured in any documents. Corporate life has formulations no one writes down. Some are cutthroat, some are kumbaya, and most are a matrix where influence moves faster than org charts. As scientists, we assume logic runs the room, but it doesn't always. That's when you learn to translate.

I learned it the night I started my first MBA course, with our firstborn in a bouncy rocker at my feet, Bob Marley on loop, and a take-home exam open on screen. Yelena and I were both working full-time; we were new parents but not exactly new to the corporate "game." I wasn't chasing letters after my name. I was learning the language of the room: finance, claims, supply, and the "why" behind the decisions that hit the lab like unpredictable weather.

Yelena: Translation isn't selling out the science. It's connecting it, so the right people say "yes" sooner, and the work survives the meeting. I didn't realize how valuable it was to see it happen in real time.

Three Truths That Changed How We Played

- **The unwritten menu matters:** Culture is the house seasoning; it's useful when you know it, and dangerous when you don't. Learn who actually makes decisions, what "fast" means in your company, and how conflict gets handled. **Tip:** Ask how the last products launched, and it'll tell you everything you need to know.

- **Titles aren't the map; influence is:** You need to know the people whose jobs directly affect the timing of your launch, like those who ship product, the functional department that approves your formulation, the team that develops the product, and the buyer who hates surprises. Find them. Feed them context. Make them, and your boss, look good.

- **Promotions come backward:** You don't get the title and then do the work. You do the next job until your manager can't defend not giving you the title. Unfair? Often. Empowering? Yes, because you can start today.

José: The most common review-season question I've heard: "What am I doing wrong?" Usually, nothing. You're just not doing anything extra, yet. Own a problem no one wants. Deliver without drama. Build a coalition. Then bring the receipts.

Yelena: And keep your boundaries. Being reliable is not the same thing as being available to be used.

José: I remember the day it clicked. I was in a conference room with the leaders of Adelante, our Latino/Hispanic employee resource group (*adelante* means forward), meeting with our executive sponsor. Someone asked for their advice on how all the future leaders in the room could move up.

The executive didn't sugarcoat it: "Don't let requirements get in your way. Nothing is handed to you because you're here. If the role calls for a master's, go get the degree."

It was blunt. It was fair. And it was exactly what I needed. I realized my résumé, not my potential, was being measured. If I wanted the next seat, I had to earn it first—before the raise, before the title. That's why I enrolled, even when the pay wouldn't match the responsibility. Looking back, that moment was the beginning of the end for me at PepsiCo; it was the moment I stopped waiting and started moving.

Both: That advice set the tone for everything that followed: learn the rules, exceed them, and choose the moves that stretch you.

The Old Contract Is Gone. Write Your Own.

There was a time when companies traded loyalty for pensions and permanence. Now, it's 401(k) matches, sign-ons, and maybe a 1. 75% year-end raise if the moon is kind. Every time we moved, our compensation jumped double digits. Is that disloyal, or just honest math?

We have friends who stayed, happily, and built beautiful careers where they began. Loyalty isn't dead; it's conditional. The key is knowing your conditions. For us, it was simple: grow or stagnate. We chose the moves that scared us a little and stretched us a lot.

Layoffs, Reality, and What You Can Control

Layoffs aren't meritocracy; they're context. You can't eliminate the risk, but you can tilt the odds.

- **Be portable:** Collect skills, stories, and references you can carry anywhere.

- **Own the ugly problems:** Be the call people make first when they need help with quality fires, cost resets, line trials, etc.

- **Keep a living brief:** Have one page that shows your impact in numbers, not just your duties.

- **Invest in the corridors:** Build trust with suppliers, plant leads, regulatory, finance—the places where decisions congeal.

Yelena: And build a life that can absorb shock. I grew up in a minimalist house, which helped me understand that fixed costs are a superpower.

José: Part 2 was us learning the kitchen: heat, timing, politics, and people. Not just how to formulate, but how to ship. The maze didn't get simpler; we got better at reading it.

Yelena: We stopped saying "one day" and started saying "Day One." The difference showed up in how we worked and where we were willing to go next.

José: If Part 2 was learning the map, Part 3 is choosing a destination and traveling there with purpose.

About layoffs: I wish I could help you avoid what we call Level 6 on the Barbosa Resiliency Quotient (BRQ)—the layoff level (more on the BRQ later). But selection is often arbitrary, contextual, and conditional. You can prepare, adapt, and stack the odds in your favor; you can't eliminate the risk entirely.

That's the point. Corporate life isn't a straight line; it's a chessboard where the rules shift mid-game. Endurance matters, but agility wins. The people who rise after the shakeups aren't the ones clinging to the old opening; they're the ones who can pivot, create new value, and play the position they actually have.

Yelena: Part 2 taught us the openings. Part 3 is about playing the board in front of us, and when needed, flipping the board to build our own.

José: That ability to innovate on the fly, in the lab, in the room, and in your own life, isn't just survival. It's a competitive edge, and it's exactly where we're heading next.

Part 3: Side Dishes and Secret Sauce: Growth, Grit, and Grace

Chapter Eight

The Spice Rack: Innovation and Creativity

"Everything that can be invented has been invented." That little gem is often attributed to Charles Duell, the commissioner of the U.S. Patent Office in 1899, except, spoiler alert, he never said it. Still, whoever first coined the phrase had clearly never spent time in a food science lab. Innovation is endless. The opportunities to discover, refine, and reinvent are as open as a test kitchen on a Monday morning.

In our world, innovation isn't just about dreaming up the next kombucha-adaptogen-probiotic-super-drink; it's about solving the hundreds of small, maddeningly specific technical puzzles that make it possible. Switch from cane sugar to monk fruit, or my personal favorite, Reb-M, and suddenly your sweetness curve shifts, your acid perception changes, and your aroma balance takes a left turn. It's less "swap and serve" and more "Jenga tower in a windstorm."

It's no surprise that the title *Innovation Manager* attracts attention. In every job posting I've ever put up with that word in it, résumés flood in. It carries a certain prestige, a glow, that makes people want it on their LinkedIn profile. Reality rarely matches the mystique, though. The expectations are high, and true breakthroughs are rare. Fire was an innovation. The beverage was an innovation. What are we mostly doing now? Iterating: new flavors, ingredient combinations, and delivery formats. All exciting, but let's be honest, we're not splitting the atom here.

That doesn't mean the work isn't deeply valuable. It just means that true innovation is hard, expensive, and slower than quarterly earnings calls allow. This is why the savviest marketers, the best innovators, and the most valuable R&D partners all know the same truth: If you want to wear the crown of "innovator," you'd better have your R&D team on speed dial and treat them like royalty.

One of the challenges, and joys, of innovation is separating signal from noise. Scientists are naturally drawn to great experiments, whether or not they fit the company's "north star" (signal) at the moment. I've also seen "pet projects" (noise) pushed forward just because a leader had the clout to make it happen.

That's where the role of translator comes in. It's not an official title, but it should be. Translators are the ones who bridge the language gap between strategy and science, turning the business brief into experiments scientists are excited to run, and turning the lab's data into stories business leaders can't ignore. When done well, translation doesn't just keep projects moving; it elevates them.

One of my teams recently proved this in the most gratifying way. We launched a product into a category so saturated that most people would have written it off as "unwinnable." Launching successfully meant we had to translate the brief properly, execute flawlessly on the non-negotiables, and still find the space to slip in quiet moments of product mastery, like a delicate lemon oil coating that made the experience just a little more special. The result? The product took home a Nexty award at the Expo West show in Anaheim and, of course, contributed to the earnings results of the company.

That award celebrated more than just the product. It honored a team that understood each other across functions, with translators ensuring nothing was lost in handoffs. Watching them bring that product to life, knowing they had navigated the science, the strategy, and the subtle artistry, was one of the proudest moments of my leadership career.

Over my career, I've helped launch hundreds of products. Add Yelena's, and together we've put more items on shelves than some brands do in their lifetime. The ones that endure tend to be the ones where translation does the heavy lifting, where we translate the chef's gold standard and make it work for supply chain, regulation, and a Tuesday night shopper.

There's nothing like walking into a store, spotting something you developed, and watching a stranger put it in their cart. The five-star reviews, the texts from friends—those moments never get old.

Here's the hard truth most of us would rather duck: Not every launch hits. Sometimes the product you'd bet your 401(k) on fizzles. Consumers are human; research is a flashlight, not a guarantee.

Case in point: Years ago, a countertop cold-beverage appliance performed like a dream in pre-launch testing. In concept rooms, people beamed; the prototypes drew crowds. But once it hit the market, reality struck. Its price, footprint, noise, use-case friction, and sales landed far below projections. The lesson wasn't "don't innovate." It was "weigh the real-world frictions early, then iterate."

This brings me to my favorite truth about innovation: Iteration is the backbone. The most iconic products don't debut perfectly; they earn it, version by version, use-case by use-case. First, Oreo launched. Then came Double Stuf. Then Mega Stuf. You don't delay the first Oreo until you can make Mega Stuf; you ship, learn, and refine. I've watched organizations sabotage themselves, waiting for perfect and letting "good enough to launch" die on a vine.

Pro tip: In food science, as in life, you must iterate your way to greatness. Launch. Learn. Refine. Repeat.

Yelena: Translation gets you on the shelf. Iteration keeps you there.

Day One starts with the version you're brave enough to ship.

GMO: Fear, Facts, and Food Science

Innovation is about pushing boundaries, but pushing boundaries can make people nervous. The more technically advanced or mis-understood a technology is, the more resistance it tends to face.

Take genetically modified organisms (GMOs), for example. Few topics in food science create more heated dinner-table debates. Much like mRNA vaccine technology, the science behind GMOs is often overshadowed by fear, politics, and headlines that read like they were written by a sci-fi screenwriter.

GMOs aren't magic bullets, but they're also not the corporate monsters some people make them out to be. At their core, they're simply a more precise way of doing what farmers and plant breeders have been doing for centuries: making crops better. The difference is precision and speed.

The Upside?

- **Nutrition, boosted:** Certain GMO crops can be designed to contain higher levels of vitamins and minerals than they naturally would, giving us a fighting chance against nutrient deficiencies. Soil degeneration has a worthy opponent.

- **Resilience:** Built-in drought, pest, and disease resistance can mean healthier crops with fewer pesticides.

- **Food security:** Higher yields from the same farmland can help feed more people without pushing deeper into fragile ecosystems.

It's worth remembering that over the past century, soil erosion, changing climates, and modern farming practices have chipped away at the natural nutrient density of crops. GMO technology can, in some cases, reverse that, giving us produce that's not "just as good as it used to be," but even *better*.

The Risks?

- **Corporate control of seeds:** Patents and licensing agreements can concentrate too much power in a few companies.

- **Biodiversity:** If GMO crops dominate planting, we risk narrowing our genetic diversity, making food systems more vulnerable to shocks.

- **Trust:** No technology works if the public doesn't feel it's transparent, safe, and well-regulated.

Like any tool, GMO technology is neither inherently good nor bad; its impact depends on the intent behind it and how it's applied. We've seen it deliver remarkable benefits: improving crop resilience, enhancing nutritional profiles, and potentially helping close the gap in global food insecurity. Yes, we've also seen legitimate concerns: corporate misuse, over-reliance on monocultures, and the need for stronger oversight.

The point isn't to blindly cheer or categorically reject it. Instead, we should ask: What's the purpose behind innovation? When that purpose is rooted in the benefit of all—nourishing people, protecting resources, and making our food supply more resilient—it's a purpose worth pursuing.

Fear often comes from complexity, but so does progress. The role of scientists, and increasingly AI-assisted science, is to harness that complexity for the common good. If we shy away from innovation out of fear alone, we risk losing the chance to solve some of our most pressing challenges. If we lead with purpose, we can make those innovations work for humanity, not against it.

Purpose isn't just a question for science or technology. It's also deeply personal. It shapes how we show up at work, how we navigate the unwritten rules of corporate life, and how we balance ambition with the rest of our lives. Nowhere is that more evident than in the experience of women in corporate roles, where the lab coat is just one of many hats worn in a single day. For many, that calculus is uniquely relentless. In the next chapter, we'll explore these stories.

Chapter Nine

Family Style: Women in Corporate Roles, Parenthood, and Life Beyond Nine-to-Five

Women in Corporate Roles

Yelena: It took many years of being a bystander to understand how to navigate the corporate world, even with José providing insights along the way. In the early months after graduate school, during my first corporate job, I loved my personal style: high heels, pencil skirts, and fashionable tops that let my personality shine. I knew it wasn't exactly proper lab attire, but it was me. So, like any problem-solving scientist, I found a workaround: I stashed a pair of flats and safety shoes under my desk and would switch shoes multiple times a day. Heels in the hallways, steel-toes in the lab.

A few months in, I was walking down the hallway when a higher-level executive, not someone I reported to directly but high enough in the ranks, glanced me up and down and said, "Oh, you must not be working in the lab today." I gave a polite smile, we passed each other, and that was that. Later in the day, I bumped into their administrative assistant and casually mentioned the exchange. This admin was someone I admired, seasoned in the industry, sharp, and incredibly stylish. They paused and said something that stuck with me: "Your appearance builds your reputation before your reputation has a chance to build itself."

That was all I needed to hear. From that day forward, the pretty heels stayed home. Not because I feared leadership or tried to conform, but because I wanted to earn respect for my science first. Once that credibility was in place, then I could bring the full spectrum of my personality, including my wardrobe, into the spotlight.

Ever since, at every company I've worked at, I've led with substance first. That doesn't mean I stopped wearing colors or completely abandoned my style; it just means I became more intentional about how I showed up, knowing that my presence often spoke before I got the chance to.

José: Early in my management career, I had an experience that taught me an important lesson about intent versus outcome, and it came courtesy of my wife.

One of my team members had just returned from maternity leave. Before she left, she had expressed interest in attending an upcoming industry conference. I was deciding who from the team would go, and I found myself talking it through aloud with Yelena at home: "I'll probably send others this time. She just got back, and a conference would be too hard on her."

My intent was pure. I thought I was protecting her from unnecessary travel stress right after coming back to work. Yelena's reaction? Let's just say she was... less than impressed. She immediately cut in: "You cannot take her decision out of the mix. If she decides she can't go, fine, then choose someone else. But deciding *for* her is wrong."

Whoa. That stopped me in my tracks. She was right. My intent didn't matter if my actions took away someone's choice. Even though I thought I was being considerate, I was undermining her autonomy.

That moment changed the way I think about leadership. People, especially women in corporate environments, don't need decisions made for them under the guise of protection. They need the space to decide for themselves, with the same opportunities as everyone else. My job wasn't to pre-filter those opportunities; it was to make sure they were visible, accessible, and supported.

In case you're wondering, she went to the conference. She crushed it.

That experience also got me thinking more broadly about gender and leadership, not just as it related to my own decisions, but to people I've worked for over the years. Across my career, I've had just about an even split between male and female managers, and when I tally up the good versus the bad, it's an exact 50/50. So, from my experience, gender is ruled out as a reliable metric for whether someone will be a good boss. The only real differences I've noticed are in how we communicate and how I'm perceived. My male managers have typically been more concise in one-on-one meetings, using succinct language and keeping the conversation strictly work-focused. Unless, of course, the topic turned to sports. In that case, those "short" conversations suddenly ran long—way longer than any personal-life tangent I've ever had with a female boss. My female managers, on the other hand, have tended to be more verbose overall, with discussions that sometimes drift into personal life.

I've also noticed one behavioral pattern that's never happened with a male boss but has with a few female managers, even some of the best ones I've had. Occasionally, they'd show protective tendencies that felt... well, almost maternal. Not in a bad way. In fact, it usually came from a place of genuine care. But it was different, and I had to get used to it. I've never had a male boss try to "dad" me in that way.

It's not criticism, just an observation. These differences don't make one better or worse; they simply shape the texture of the working relationship. For me, the most valuable takeaway has been this: Great leadership transcends gender. It's built on respect, trust, and clarity, and those come in many styles.

Leadership skills don't stay neatly contained within office walls. The same adaptability, empathy, and creative thinking you use at work inevitably spill over into life outside of it. Which brings us to a different kind of proving ground for innovation...

Innovating on the Home Front

José: Innovation doesn't stop at the lab bench or the boardroom. In fact, some of our most creative problem-solving happens in the least glamorous of settings, like in the kitchen at 6:30 a.m. while negotiating cereal preferences, show locations, and whether socks are really that important. Honorable mention goes to shower time. It's the perfect reflection spot, but I digress.

As professionals, food scientists are trained to think in iterations: pilot, pivot, and refine. These same instincts are vital at home. Parenthood, especially when combined with demanding careers, requires a kind of relentless innovation. You build systems, only to outgrow them. You master routines, only to have them shattered by teething, spelling tests, or the daycare stomach bug.

Recently, our kids were watching Boss Baby, and one moment stopped us in our tracks. The pint-sized CEO tells his team they need to give 100% to the mission and 100% to their families. One of the babies shouts, "Yellow 100" with total confidence, as if this makes perfect sense.

Honestly? It kind of does.

That moment, ridiculous as it was, captured the daily math-defying logic of working parenthood. You give everything you've got at work, and then somehow give everything again at home. The percentages may be impossible, but the intent is real.

No, we don't use "Yellow 100" as a mantra in our household. But the heart of the idea resonates. Parenting while working isn't a formula you solve once. It's an ongoing exercise in adaptation, creativity, and grace. It's the ultimate innovation lab.

Yelena: Motherhood has taught me that multitasking is much more than a skill; it's a superpower. A friend recently reminded me of the first time we met, recalling how she spotted me at a playground, simultaneously playing with my toddler, nursing an infant, and conducting what sounded suspiciously like a job interview, all without missing a beat. Only perspiration gave away the weight of this juggling act. She was in awe, but for me, it was simply doing what needed to be done. That phone call turned out to be an interview for a role at Welch's that would become a pivotal step in my food science career. It's just one example from a lifetime of moments where being a mom and a professional means juggling countless responsibilities and somehow managing to thrive, not just survive.

My kids play with their food. While other parents stress out because their kids use their food like a painter uses a canvas, I encourage our kids to touch, feel, and experience food in ways that leave a mess only a loving husband can clean afterwards (thanks, José!). Having our kids participate in food preparation is slow, tedious, messy, and yet so important for their understanding of food preparation and their experience with food. We believe their participation encourages them to taste the food, to have fun with the food, and maybe someday, to appreciate how messy things can get.

Pro tip: Let kids play with their food, and don't cry over spilled milk

José: Speaking of multitasking, many experts say it's not possible to multitask and do everything well or even effectively. I tend to disagree, at least with the concept, not the definition or the science of multitasking. Scientifically speaking, they're correct: Our brains cannot truly multitask. However, in real life, what passes as multitasking is not only doable; it's a skill that many of us have mastered in order to, as Rudyard Kipling wrote, "fill the unforgiving minute with sixty seconds' worth of distance run."

Doing many tasks in the same span of time requires a certain level of expertise in each one, plus the ability to plan before executing. That's what allows you to switch between tasks seamlessly and still keep them all moving forward. Gary Keller, in *The One Thing*, calls this "task switching," which I'd agree is more accurate than "multitasking." My one nuance? For all these different tasks you're bouncing between to be effective, each one needs to be moved to a point of significance, or finished altogether, before you shift your attention.

If you're writing a book and taking care of kids, the first priority is making sure the kids are safe. Otherwise, the irony of losing one while writing a chapter on parenthood would be... palpable. If they're fine, though, you can rotate between writing, cooking dinner, doing the laundry, and actually finishing that chapter. In the end, multiple tasks are completed in an effective way, regardless of how you define multitasking.

In the corporate world, I've seen leaders set "no multitasking" rules for meetings. Most people would agree that this is prudent and keeps people engaged. I, for one, used to detest it because I felt it removed one of my superpowers. Yes, I can email and listen to the message of a meeting at the same time. Do I miss minor details? Sure. Just as speed readers don't read every single word yet still absorb the main story, I can do the same with presentations. My true kryptonite? Presenters who read their slides aloud; I've already read them twice.

Pro tip: Multitasking is a responsibility. If you can handle it and have mastered it, go for it. Otherwise, the perils are many, so approach at your own risk.

As a parent, especially of more than two kids, multitasking isn't optional—it's survival. Man-to-man defense works fine until you're forced into zone defense against multiple little forces of nature. We have four kids; that's two per parent at best. Sometimes one of us juggles all four while the other is consumed with guilt for leaving their partner to fend off the adorable but ruthless Gremlins™ we love so much.

Managing a work-life blend is like balancing beakers and bedtime stories; both can blow up in your face if you don't pay attention. In fact, managing four kids is a lot like managing an emulsion: You need constant energy input to keep everything suspended, and the second you stop paying attention, things start to separate.

Opportunity Costs: The Hidden Equation of Parenthood

Parenthood forces you to become fluent in opportunity cost, whether you've read an economics textbook or not. Every decision comes with trade-offs. If we go to the school fundraiser tonight, we won't be home for dinner together. If we take the big work project with travel, we're missing sports games and bedtime routines. If we skip that networking event, we might miss the connection that opens a future door.

In corporate life, opportunity cost often hides in job decisions. Staying in a role might feel comfortable, but if another opportunity offers more growth or better alignment with your values, the cost of staying becomes very real, even if you don't see it on a spreadsheet. The same principle applies at home. Choosing to socialize on a rare free weekend might mean less time to reset for the week ahead, but if that gathering fills your emotional tank, maybe it's worth it.

We've learned it's not always about avoiding high-cost choices. Sometimes we intentionally choose them. We've said "yes" to hosting a dozen kids for a birthday party when it would've been easier to go low-key. We've taken on passion projects knowing they'd eat into our downtime. When we make those choices consciously, the stress changes shape, becoming purpose-driven rather than draining.

It's like formulating a new beverage: You can't optimize every variable at once. Boost the sweetness, and you'll need to rebalance acidity. Add protein, and you'll have to manage viscosity and texture. Every change is a trade-off, but if you make them intentionally, you end up with something you're proud to put your name on.

Parenthood works the same way. Every "yes" comes with a "no", but if the "yes" aligns with what matters most, the "no" feels less like a loss and more like an investment.

Mini Lab Coats, Major Curiosity

Yelena: I don't think I ever intentionally sat down and reflected on how our relationship with food connects to the way we're raising our children. When I look back, it's clear that from a very young age, food has been both a learning tool and a playground in our home. With each of our kids, there's been a phase where we'd plop them on the kitchen counter, pull out the spice rack, and just explore. We'd sniff, occasionally taste, and mostly just talk about what they were experiencing. Vanilla and cinnamon always became early favorites. A whiff of freshly ground pepper usually ends in a sneeze—snot-filled, but memorable to any parent whose shirt is permanently stained with kid's snot (check the upper shoulder area for a telltale sign the person you are speaking to is a parent of a young child).

The unspoken lesson? It's not only okay to play with your food but also encouraged. We've never pushed our kids toward food science as a career (though we wouldn't complain), but we've always believed that mealtimes should be participatory, playful, and curiosity-driven. Sure, we've got a few picky eaters, but this approach has taught our kids not to fear the unfamiliar. Whether that means watching from a distance or helping prep dinner themselves, they're always being invited into the experience.

We are their role models: two grown adults who still play with their food for a living and love every minute of it. I hope that one day, when our kids are grown up and in their own kitchens, they'll remember the messes, the smells, and the laughter. Professional or not, we're all a little bit of a food scientist at heart. Whether you're layering pickles on your tuna salad (don't knock it until you've tried it), creating a new sauce, or shaking up a cocktail, you're experimenting. You're creating. You're playing. And we're all in full support of that.

José: One of the things I'm most proud of as a food scientist isn't on my résumé; it's the fact that I brought my kids to the lab whenever I could. It wasn't a publicity stunt or a bring-your-kids-to-work day; it was just life. If I were swinging by the lab on a weekend or off-hours, I'd grab their car seats and a bag of snacks, and off we would go. Some families go to theme parks for fun; we went to the bench to see beakers and hear the purring of the machines in hibernation, waiting to come to life during work hours.

Let me tell you—they loved it. Not necessarily because they were mesmerized by the science (though some days, I could see their little gears turning), but because the lab was basically a free vending machine of corporate swag. Stress balls, branded pens, sticky notes—it was like trick-or-treating in a world powered by chromatography. I once caught one of them walking out with a notebook and a foam orange in each hand, grinning like they'd just robbed a stationery store.

To them, I didn't just "work in food." I worked in magic or pizza, as they would answer when someone asked what I did for a living. They saw soda fountains that didn't belong in restaurants, gummy prototypes no one else had tasted, and a place where grown-ups played with beakers instead of laptops. More importantly, they saw that work didn't have to be something you dreaded—it could be something you shared.

Those moments reminded me that careers aren't just built in offices or labs. They're built in the spaces in between, when you're showing your kids what passion looks like and giving them the permission to imagine themselves in your shoes (preferably steel-toe, safety-approved shoes).

Yelena: Motherhood isn't something you can ever be fully ready for, and I say that as someone who decided to do it four times while building a career in food science. Four kids. Two working parents. No live-in help. Just us, our coffee, and a calendar that looks like an air traffic control board.

To people outside our circle, four kids sounds like chaos. Inside our home, it still is, just chosen chaos. We knew the math: more mouths, more calendars, more feelings. We also knew the payoff:

a big, loud family that would teach us more about leadership, patience, and resilience than any boardroom.

For most of our lives, my kids have only known me as a working mom. Yet at the beginning of 2025, I made a conscious decision to stay home with them. As you can guess, they loved it. I've loved it too, even with the endless, "Mom, can you...?" moments and the logistical act of managing four very different lives under one roof. I've swapped product timelines for orthodontist appointments, and strategic planning decks for reading logs and school curricula. Yet, the same part of me that could deliver a flawless product launch now thrives in delivering a smoothly run day for the family.

It's not without its tradeoffs. My social life often comes last on the list, not because I don't value my friendships, but because after juggling kids' schedules, life's deadlines (like for this book), and day-to-day operations of our home, the thought of dressing up for a night out feels like prepping for a second shift. Social media helps fill the gap; quick check-ins, shared memes, and voice messages at odd hours let me stay connected without leaving the house. It's not the same as lingering over coffee with a friend, but it works for now.

Speaking of social life, before kids, José and I lived in what you could only call a global village. Our circle of friends spanned countries, from Spain to South Africa, and every time zone in the United States. Our wedding guest list looked like a miniature United Nations: friends from across the globe alongside family from Russia, Puerto Rico, and Israel. The dance floor was a joyful collision of accents, languages, and styles—salsa blending with hora, hip-hop sharing space with Danza Kuduro... and our friend Lance trying to stand up and give the best man toast in a wedding with no best man or bridesmaid.

Back then, dinner parties were as much a science experiment as a social event. I was deep into exploring molecular gastronomy, fascinated by foams, spheres, and liquid nitrogen ice cream—not because I had to for work, but because I wanted to see how far food could be pushed before it became art. Friends were willing guinea pigs, and our evenings often blurred the line between a lab session and a feast.

Those friendships shaped us as much as any career milestone. They expanded our view of the world, taught us cultural fluency, and made us feel at home almost anywhere. Four kids later, the social calendar shifted. Dinners out became dinners in, not the candlelit kind, but the "pass the chicken nuggets while negotiating broccoli consumption" kind. The vibrant, globe-spanning social life is still there, but now it's kept alive through group chats, social media, and the occasional late-night video call when time zones and bedtimes align.

Recently, our son asked why the queen is more powerful than the knight in chess. I told him: "The queen has range. She goes where she is needed. She makes everything around her stronger." That's how I see motherhood. The knight is noble, but it moves in L-shapes, while mothers cover the whole board. We protect, we advance, we adapt. When you're raising four kids while steering a career or a household, you learn fast. You can't just move in patterns. You have to own the board.

And at our house, the chessboard sits next to the beaker.

As any good formulator knows, keeping an emulsion from separating isn't about one ingredient; it requires balance, timing, and constant, subtle adjustments. Parenthood works the same way. One day it's homework, the next it's a fever, and all the while you're keeping the long recipe in mind: a safe, curious, happy home. Some days the mixture is silky; other days it's close to breaking. The secret is the steady hand and the daily, loving stir that holds it together.

Shortly after having our first child, we were living in Vermont, juggling two demanding full-time roles. Because José doesn't believe in downtime, he decided it was the perfect moment to start his MBA. On top of it all, we were learning how to be parents. Turns out, when you leave the hospital with a newborn, they don't send you home with a Parenting 101 handbook. We were lucky to work for a company that supported working parents and had managers who understood the importance of flexibility. Even better, we stumbled into a nanny share with a colleague who lived just down the road from the office. With our firstborn safe in a harbor a minute away, the rest of the day felt navigable.

Those early months were a complete blur. I was managing fast-paced projects across two locations and trying to decipher whether a baby's cough was just a cough or something more serious, all while figuring out how to build a pumping schedule into an already hectic calendar. (Too much information or not, I became an expert at pumping between meetings and even while driving between sites. Safety first, but efficiency always.)

By the time we had our second, then third, then fourth, the rhythm of our life was less about balance and more about fluidity. We learned to readjust constantly, reimagine our schedules, and redefine what "normal" looked like. Through it all, we kept showing up for our family and for our work because both mattered deeply. Somehow, amidst the chaos, the sleep deprivation, and the growing laundry piles, we still managed to crush it—or maybe we were the ones getting crushed. It's all a blur.

Sidebar: For the Pet Parents

Food science doesn't just stock your fridge; it fills your pet's bowl too. There's a whole sub-world of scientists who lose sleep over kibble crunch, tuna aroma stability, and whether dogs really prefer peanut butter or just tolerate it because it hides the pill.

I've never worked on pet food, but my brain can't help itself: I see a bag of kittles and immediately start mentally reverse engineering it like I'm reading the ingredient list on a gourmet snack.

"Chicken by-product meal" reads legal; used well, it's simply a concentrated, organ-based protein. "Natural flavor" is a tiny dose of flavor and aroma compounds from natural sources that helps with palatability. It serves as proof that "taste is king," even in the pet world, since it's what makes Mr. Whiskers purr. They're options, not morals. If the brief calls for "no natural flavors," we switch palettes—broths, rendered fats, yeasts, gentle browning, fermentation—so pets still choose the bowl, and you still like the label.

Like nutritional and supplement labels in the human world, pet food labels are a fascinating mix of marketing poetry and scientific reality. "With garden vegetables" means there's probably one dried carrot fleck somewhere in the mix, the way a confetti cannon leaves a single piece of paper stuck in your show after a parade. "Grain-free" often means "we swapped corn for peas," which is like replacing your soda with a milkshake—still calories, just different pants.

The feeding guidelines? They're the "serving suggestions" of the animal kingdom, and optimistic at best. Your dog either inhales it in 45 seconds or turns into a canine restaurant critic, sniffing and walking away until you add a drizzle of something fancy. In the end, it's still food science, balancing nutrition, shelf life, texture, and taste (for someone with a completely different taste bud map). It's just that in this lab, the most important sensory panelists can't talk back. Well, they can, but it's usually barking.

While I never made pet food, I did grow up with what could only be described as a small but eclectic zoo: dogs, cats, ferrets, rabbits, iguanas... and, for reasons I still can't explain, a large white crab. The ferrets were mischievous, the iguana aloof, and the crab... well, the crab mostly minded its own business, which is more than I can say for the rabbits.

Each species taught me something about responsibility, chaos management, and the limits of what you can train with food incentives. Once I had kids, I realized my pet tolerance had been spent in advance. I traded litter boxes and aquariums for diaper pails and Lego mines. I still salute anyone juggling both kids and pets because it's a level of multitasking that deserves its own Olympic category.

Maybe that's why I've always admired storytellers who capture chaos without making it feel chaotic. The ones who distill love and loss into something still, something warm.

Years ago, I stood in line to meet Nicholas Sparks, not once, but twice. I was one of the only men in a sea of fans, and maybe that's why he pulled me aside and asked me to stay and chat while he signed. We talked more than we should have. He told me about how he used to have a Puerto Rican girlfriend; I told him stories that probably made no sense out of context. For a few moments, it felt like I was talking to someone who saw life the same way as I did: messy, tender, and worth writing about anyway.

As Nicholas Sparks writes in *Nights in Rodanthe*, "Before we met, I was as lost as a person could be, and yet you saw something in me that somehow gave me direction again." That line felt like a mirror, revealing who I was in the few years leading up to when I met Yelena. Lost, then slowly finding my way.

Yelena: I didn't have a map either, but together we decided that forward counted.

José: This next section is my tribute to that feeling, to Nicholas Sparks's quiet intimacy, and to the way love can turn one day into Day One.

The Original High-Performing Team: Inspired by a Conversation With Nicholas Sparks

There's something curious about love when it's been tempered by chaos. It no longer needs violin soundtracks, dramatic rainstorms, grand gestures, or cinematic scripts. It lives in quiet places, half-finished sentences, knowing glances across a chaotic room, and the steady rhythm of two people choosing each other over and over again.

Love, for us, shows up in myriad ways. It's in tag-teamed bedtime routines, grocery runs, and simple reminders to thaw the chicken. We're not roommates. We're not each other's boss. We're definitely not playing the tired "who-does-more" game, though I have a feeling I know who'd win that argument.

What we are is a high-performing team. The kind of team corporate executives spend millions trying to build: minimal ego, maximum humor, and a shared mission that matters more than any single task. In our case, the mission is simple: raise four kids, stay happily married, and try not to run out of milk... again.

We've made a habit of stepping in wherever the ball might drop. If Yelena's buried in work, I'm on dinner duty. If I'm stuck on a call, she wrangles the bedtime routine chaos. There's no scorecard—just a silent understanding that whoever can, will. When one of us is gone, even for a night, the silence is louder. The absence of that invisible glue becomes palpable. Even the kids sense it and modify their behavior in an obvious way.

However, it's not all logistics. Our story breathes a softness, too, like the way we still reach for each other at night, legs tangled, breath slowed, neither one of us needing to say a word (this part feels like Sparks, right?).

Almost every night, one of us whispers, "It's the best feeling in the world." On the nights we don't, it's only because we passed out mid-whisper.

We didn't build this overnight, but we did build it. Every day, we keep choosing to build it again because what we have is more than love. It's infrastructure. Whether it's in marriage or management, resilience is the through line. The soft kind that shows up in whispered goodnights, and the gritty kind that shows up in office politics, surprise re-orgs, and PowerPoints with 67 slides too many.

Love teaches you a lot. Corporate life? That's a whole different curriculum.

Chapter Ten

The Heat in the Kitchen: The Barbosa Resiliency Quotient (BRQ)

A Corporate Survival Scale for the Bold and the Brave

L et's face it, most of us spend more time at work than we do with our families, our hobbies, or our beds. Ideally, those hours should be mentally stimulating, emotionally fulfilling, and, dare I say it, fun. For many, though, work feels more like SEAL training run by an instructor who cc's the legal department: a combination of sprints, cold plunges, and a mandatory webinar on email etiquette.

People often ask me, "How are you still smiling?" Truth is, I've taken my L's. I've missed lab promotions, survived lab meltdowns, and received end-of-year reviews that read like lukewarm Yelp posts. I've learned that a positive mindset is like fuel for the "fire in the belly;" it gets you through the hard days and oftentimes sets you apart.

Some days, resilience means channeling your inner Rocky Balboa while "No Easy Way Out" blares in your head, and refusing to let one more lab disaster, bad end-of-year review, or skipped promotion knock you out of the ring. Resilience is a precursor to adaptability, and the ability to adapt is well known in scientific circles as the Darwin rule. If Darwin taught us anything, it's that it's not the strongest who survive, but the most adaptable. In corporate life, that truth is gospel. In our world, resilience is both a mental and technical skill. Ingredients get banned? Reformulate without losing taste or texture. Supplier goes under? Re-source and re-validate without delaying launch. Adaptability is survival, for products and for people.

To help put this into context, Yelena and I, lifelong food scientists and self-appointed resilience anthropologists, have cooked up our own version of the Schmidt Pain Index (yes, the one dreamed up by entomologist Justin Schmidt and made internet-famous by Coyote Peterson's bug sting escapades). Instead of bullet ants and tarantula hawks, our index measures corporate and life curveballs that test your grit, patience, and sense of humor. Except ours is less venomous (mostly), more relatable, and aimed at helping you grow, not just survive.

Introducing...

The Barbosa Resiliency Quotient (BRQ™)

A lighthearted, yet frighteningly accurate, diagnostic tool to measure your ability to take a career punch to the gut and keep moving forward.

This isn't meant to scare you (okay, maybe a little). It's here to help you prepare, like a good Boy Scout or a parent before back-to-school shopping. Think of it as your psychological weight-training for when the career barbell gets too heavy to lift alone.

Use the following scale to test how resilient you currently are to career curve balls. Each scenario is rated for the level of resilience it demands. For example, "Resigning with no next job lined up" has a Barbosa Resiliency Quotient of 8.5; it demands courage but can be a powerful career pivot and embody the ultimate growth behavior. As you read, notice which experiences you've faced or feel ready to face. The higher your BRQ™ score, the more resiliency it builds, although it comes with a "tough experience" tradeoff.

Level 0.5 – The Mandatory Ice Breaker

That awkward moment in a meeting when someone says, "Let's go around and say two truths and a lie!" You feel a strong urge to fake a Wi-Fi outage. Your palms start sweating. This is low-stakes discomfort training. It reminds you of an aftertaste you dislike, but you've already taken a huge gulp—you'll survive, but your sinuses may never forget. Mild discomfort, but the necessary first rep in your BRQ training.

Resilience check: If you break out into a sweat at the words "Ice Breaker," you may want to start working on your resiliency.

Pro tip: If you're the one picking the ice breaker, make it "Tell us your best dad joke." It's harder to be nervous when everyone's laughing with you (or at least groaning in unison).

Level 1.0 – The Full-Day Town Hall

No tasks, no decisions, just listening. Lots of it. The first half-hour feels fine, but by the third hour, you're wondering if you can use PTO retroactively. Bonus BRQ™ points if you voluntarily sit in the front row, where eye contact is constant and escape routes are limited. I've posted on social media about how I encourage future leaders to sit in the front row or as close to the front row as possible during these events. To me, this demonstrates a sense of vulnerability that translates to leadership. You may be called on to ask a question, you may get more eye contact from the speakers, and yes, you may sweat a bit more sitting in front. Many responses on social media did not agree with my suggestion, which is okay.

Resilience check: If you can't find yourself courageous enough to sit in front, at least occasionally, your BRQ™ score may need a boost (and you may find the next resilience levels to be a little less to your liking).

Level 2.0 – The Dreaded Reorg

You now report to someone whose name sounds vaguely familiar, and you're on a team with a mission statement that sounds like a TED Talk title. But panic fades quickly once you realize you still have a job. You may go home a bit discombobulated, but your kids, an after-work drink, or a lengthy commute quickly distract you enough to forget about it by the next day, when it no longer stings anymore.

Resilience check: Your adaptability level is like a banana smoothie with a vitamin premix: thick but digestible.

Level 3.0 – The Mid-Year Review

Your manager says, "Keep doing what you're doing." Somehow, that both comforts and unnerves you. You replay the meeting later like it's a cryptic voicemail. Surviving this with your confidence intact builds real BRQ™ strength. No praise, no complaints, just enough ambiguity to make you question life choices. In our experience, even when we were getting good reviews, the constructive criticism that came with them always felt stronger and more important than all the positive signals put together.

Resilience check: If you can survive the mid-year review, you're well on your way to handling the next career curveball.

Level 4.0 – The Beloved Boss Leaves

It starts with a vague calendar invite: "1:1 Catch-Up." You already know what's coming. Your stomach drops, and your palms sweat, the same way they did the first time you opened the R&D Kitchen door to taste the ghost pepper edition chips you helped develop. You haven't even taken a bite, but your body already knows it's going to burn. Having a great boss is the epitome of a great job, so when they tell you that they're leaving, it feels less like a personnel change and more like a tectonic shift. You thank them, fake a smile, and cry later (or during, no judgment).

Resilience check: Transitioning to a new manager can feel like trading in your seasoned GPS for a toddler with a crayon map. The route still exists... but you're about to take some very strange turns.

Level 5.0 – The Unpaid Promotion

Surprise, they're not backfilling the open role! You're doing two jobs now. Your title remains the same; so does your paycheck. It's the career equivalent of dipping shrimp into horseradish, thinking it's tartar sauce, which, yes, I (José) actually did at a customer event. I had tears in my eyes trying to choke it down while the client, who had just hired a few Ohio State grads who knew Yelena, watched to see what kind of man had married their old classmate. I smiled, swallowed, and pretended it was fine. The same way you'll pretend it's fine when management decides not to backfill and hands you the extra workload.

Resilience check: If you can handle this extra workload without falling apart, your BRQ™ score is on the rise

Level 6.0 – The Layoff (Not You... This Time)

Everyone gets jittery. You check your phone for that dreaded calendar block. The rumors are true, but you survived. Cue the survivor's guilt, the rapid LinkedIn updates, and the awkward hallway nods with colleagues packing their desks. It feels like the boat might have hit an iceberg (lettuce?), and while you're standing on deck, you can't shake the thought that it might be time to grab a life vest. In food science terms, it's like discovering the production line next to yours just got shut down for a major contaminant issue: You're still running, but you can't ignore the smell in the air. Is it a lucky beak, or a warning shot?

Resilience check: Either way, if you've made it this far, your BRQ™ boost is unlocked.

Level 6.5 – The PIP (Performance Improvement Plan)

The corporate kiss of death. Weekly check-ins, goal sheets, endless documentation. It's like being handed a recipe for redemption, but in invisible ink. Escaping this fate builds rare, legendary BRQ™ strength. Like being told the ice cream that was served in front of you is not for you: "My mistake, sorry." The grind needed to get through a PIP is brutal. Is it worth it? Most days, I'd say no, except I've watched someone summit from a PIP, so it's climbable. If you choose to climb, choose it on purpose: get the criteria in writing, stack visible wins weekly, keep receipts, and recruit a sponsor. Once you make it to the ridge, your BRQ keeps the altitude forever.

Resilience check: You're getting up there—resilient, but a few more steps to go.

Level 7.0 – The BRQ Gut Punch: Letting Go of a Mentor

This one stings, and not in the "paper cut" sense. This is the kind of sting that makes your chest tighten. It's not just business—it's personal. You're not simply making a headcount adjustment; you're handing over your sword to the very person who taught you how to wield it.

If you can retain a real relationship post-layoff, you're basically an emotional MacGyver, somehow building a bridge out of paperclips, bubble gum, and a mutual respect that refuses to die. It's like swallowing a large pill with no water: It sticks in your throat, and it leaves a hollow space you can't quite fill.

I've been there. I didn't handle it gracefully. I stumbled through the words, second-guessed myself for months, and replayed the scene in my head like a bad movie I couldn't stop watching. Somehow, thanks to social media and the occasional birthday reminder, I held on to a thread of connection.

Some leaders have the uncanny ability to make a farewell sound like the best thing that could happen to that person. They frame it as a celebration of a career well-lived and a new chapter about to begin. I respect that skill deeply, but I just can't bring myself to paint that kind of picture. The truth is, mentors are wired to see you succeed, solve your problems, and push you to grow—and here you are, looking into their eyes, saying, "It's time to go."

They'll never fully know how much you fought for them behind closed doors. How many times you suggested 20 different options. How the business's rigid realities closed in until it wasn't just their departure, but it was, in a sense, yours too.

Resilience check: For those who've been through it, you know: This BRQ™ is almost off the charts. The sweet spot doesn't exist here. It's a 10 out of 10 in emotional difficulty. It's one of the few leadership moments that will leave you both proud of your empathy and quietly haunted for the rest of your career.

Level 7.5 – The Layoff (It's You This Time)

Like eating a stale chip that makes you cough, it's unpleasant, un-expected, and leaves a weird taste in your mouth. You're upset, confused, and asking, "Why me?" The "Open to Work" banner feels like walking into a networking event wearing a neon sign that says *Recently Rejected*. Once the shock wears off, you realize you'll get severance and maybe even recruiter assistance. But here's the part people skip: This is also the moment to call in your network, before you disappear into a job board black hole. Reach out to people you've worked with. Share your wins, not just your availability. Use the time to update your skills, refresh your personal "flavor profile," and get clear on what you want next, not just who's hiring.

Resilience check: The job loss isn't the end of your recipe; it's the part where you get to add new ingredients to build your BRQ™ score.

Post-Layoff Survival Recipe

Prep Time: 48 hours to feel all the feelings
  Cook Time: 1–3 months, depending on market conditions
  Servings: You (but your network will get a taste.)
  Ingredients:

- 1 strong coffee (or tea, or whiskey—we don't judge)

- a dash of deep breaths

- 2 cups of updated résumé & LinkedIn profile

- 3 heaping spoons of specific career wins (metrics included)

- a pinch of humility, a pinch of swagger

- unlimited servings of networking calls

Method:

1. **Sit with it:** Let yourself grieve for a day or two—no "to-do" lists, just processing.

2. **Tell your people:** Reach out to real colleagues, mentors, and friends. Be human, not just transactional.

3. **Polish your profile:** Update your résumé, LinkedIn, and portfolio before your name drifts out of inbox memory.

4. **Re-season your pitch:** Get clear on what you want next, not just what's available.

5. **Stay visible:** Comment, post, and share ideas in your field; your next opportunity might be watching quietly.

Serving suggestion: Pair with long walks, small wins, and at least one project you've always wanted to try.

Level 8.5 – "I Quit" (No Safety Net)

You walk into your boss's office, drop the mic (figuratively... probably), and walk out with no next role lined up. There's a strange electricity in the air: half exhilaration, half terror, like stepping onto a high dive you've only ever seen from below. Your heart is pounding, not from fear exactly, but from the sheer awareness that you've set your life on a different track.

This isn't reckless. This is foresight meeting courage. You prepared when times were good, built the emergency fund, had the "we'll be fine" talks with your spouse, and learned to trust that Murphy's Law is real but survivable. You've run the numbers, and they work... at least on paper.

What now? Well, now you're in that delicious, unnerving space where the horizon feels wide open. Coffee tastes better. The air feels sharper. You're lighter without the corporate anchor, but you can feel the full weight of the world settling on your shoulders with each passing day.

Spicy and bold, this level forges leaders in pure fire. It's freedom with a clock ticking in the background, pushing you to answer questions like, "How fast can I write and publish a book?" or "How do I turn all these ideas into something that actually pays the mortgage?"

Resilience check: This leap doesn't just test your skills; it tests your nerve. Land it, and your BRQ™ score soars, leaving you with a view from midair you'll never forget.

Level 9.0 – You've Been Terminated

No plan, no warning, no severance. This is the corporate equivalent of being blindsided by a protein bar with hidden dairy on your first day of going vegan. Painful. Humbling. It's like a male subject taking too much iron—instant constipation. (Women, carry on; sometimes you do need a boost.) If you get back up, though? You're unstoppable. Neither Yelena nor I have experienced a termination, but we also haven't experienced breaking a leg bone, and according to those who have, it is the biggest pain one can endure. We believe them. I cannot fathom the resiliency either situation requires. Fired for cause means there is something that legally allows a company to remove you with no questions asked. You need to re-evaluate everything you believe is right. Soul searching is just the beginning of a long journey to understand how to get back on your proverbial feet.

Resilience check: The good news is that, once you're back on your feet, your resilience level is so high that challenges feel almost fun (almost).

Level 10.0 – The Untold Unknown

We've never met Level 10. We know its shape: no control, no map, and no clean way back. It's the door at the end of the hall that everyone swears is just a supply closet, until it isn't. It's the moment the hallway light burns out, and something breathes just out of view.

Possible Level 10.0 Scenarios:

- AI mislabeling your worth, and no human in the loop.

- A deepfake that gets believed faster than you can deny it.

- A layoff plus a non-compete that erases your options.

- Being volunteered as the culprit for a systemic failure.

- Immigration status tied to a badge that has just been deactivated.

- "Retirement" by surprise, years too early, with no runway.

If you've stood there, hand on the knob, tell us what was on the other side. Your story finishes the scale. Level 10 isn't universal; it's personal apocalypse scale.

Yelena: If you've met it, name it. Maps get safer when someone draws the monster.

José: What would you call a 10?

Opportunity Costs and the BRQ Scale

Opportunity cost is one of the most underappreciated drivers of resilience. The higher your BRQ, the more intentional you become about choosing where to spend your limited time, energy, and attention. You understand that every "yes" is also a "no," not in a scarcity mindset way, but as a strategic allocation of your resources.

People with a high BRQ aren't just good at enduring challenges; they're good at prioritizing the right ones. They don't chase every shiny project, social invitation, or career opportunity simply because it's available. They pause to weigh the cost of pursuing it against the value it will deliver. Sometimes they choose the lower-cost path to preserve bandwidth for what matters most. Other times, they consciously choose the higher-cost path because it delivers outsized impact, deep fulfillment, or once-in-a-lifetime memories.

In corporate life, this shows up in big decisions like whether to stay in a comfortable job or take a risk on a new role. In personal life, it's in deciding whether to take on a major home project during a busy work season or to block off an entire Saturday for your kid's tournament instead of catching up on email.

The real BRQ mastery comes from owning your choice. If you've consciously chosen to invest time in a high-cost, high-reward activity, you don't waste energy resenting the cost. You recognize the trade-off and lean into it fully. The resentment, the burnout, the "how did I get stuck doing this?" feeling—those creep in when you make choices passively or without clarity.

It's the same in product development. You can't have the lowest sugar, the longest shelf-life, and the richest mouthfeel without making trade-offs. Each choice shapes the end product, and if you make those choices with purpose, you'll stand behind the product when it hits the shelf.

The BRQ scale doesn't just build grit; it builds discernment. The more you are aware of your opportunity costs, the more control you have over the quality of your work, your relationships, and your life.

Final Thoughts on Fire (In the Belly)

Resiliency isn't about being unbreakable; it's about being flexible enough to bend without snapping, and smart enough to know when to pivot. So, whether you're new to the game or deep into your career, we hope the BRQ helps you laugh, reflect, and prep for whatever comes next.

And remember, when life hits hard—and it will—don't just bounce back; bounce forward. Like Rocky would say, "It's not about how hard you hit, it's about how hard you can get hit and keep moving forward." Perspective. It all comes down to perspective.

The BRQ isn't about avoiding hard times. It's meant to show how you respond to them. How you push through because there's always another BRQ level to encounter.

Hemingway said it best: "The world breaks everyone, and afterward many are strong at the broken places." That's BRQ in a nutshell. The measure of resilience isn't whether you can avoid being broken; it's whether you come back stronger exactly where the cracks once were.

In corporate life, those "broken places" might be the product launch that tanked, the dream job that turned toxic, or the micromanaging boss who made you question your sanity. Each scar can be either a warning label or a badge of honor. It's a lot like tempering chocolate. If you let it cool too fast or rush the process, it's brittle and snaps under pressure. But if you handle the break points deliberately—warming, cooling, and stabilizing—it sets into something stronger, smoother, and far more resilient than before. The same holds true for us.

Hemingway didn't stop there. He warned, "But those that will not break, it kills." Translation: If you refuse to adapt and if you cling to brittle habits and old thinking, the pressure will eventually snap you for good. Evolve or perish.

That's why high BRQ scorers bend before they break, adapt before they burn out, and see every hard season as raw material for growth. They take the feedback, the setbacks, and even the embarrassments, and turn them into steel in their spine.

For me, BRQ has always been about turning those moments into fuel. The boss who tried to micromanage me into unconditional capitulation? Fuel. The times my projects failed publicly? Fuel. The endless corporate reorganizations? Fuel. Every dent, every crack, and every disappointment is an upgrade if you let it be. See why I still smile?

Here are some tips Yelena and I put together to help us all get through these scenarios before they happen:

Resiliency Prep Tips:

- Sit in the front row at your next town hall meeting.

- Ask upper management how they got their start.

- Always keep an updated résumé (just in case).

- Learn to breathe through awkward performance reviews.

- Build strong peer relationships before you need them.

- Embrace discomfort.

Chapter Eleven

Burnt Dishes: The Importance of Failure

B oth: Let's dissect why people fear failure. As scientists, we've come across articles and experts who have written extensively on this subject. Some argue that our DNA hardwires us to fear failure; after all, failing to find the right shelter or hunt the right prey would have led to our death in prehistoric times. Others say people fear failure because it hurts our ego, and it hurts in ways that last far too long for our comfort and delicate pride.

We like to say "failure is part of success," but that's hard to re-member when your nervous system treats uncertainty like a fire alarm. Fear isn't a character flaw; it's old software. It just can't be allowed to drive.

José: When I was promoted to Director of the Innovation and Application Center at DSM, the alarm was loud. I had two years of

VMS experience in a role that bridged food and beverage with VMS, and suddenly my team jumped from 5–8 scientists to a global group of 26. An executive even dropped by for a "friendly chat" to test whether I was "too nice" for the job. Translation: Are you decisive enough to lead?

My stomach dropped. The imposter voice got busy: *You're underqualified. You'll be exposed. Play it safe.* But I didn't wait to feel ready. I moved while afraid.

What I Did (While the Doubt Was Still Talking):

- Galvanized the team to improve customer response time as well as customer engagement

- Set standards with kindness: promoted team members who modeled desired behaviors and turned "too nice" into a "clear and fair" philosophy

- Took on global responsibilities to up-level all Innovation Centers, while sharing our wins through newsletters and constant communication

Did the imposter voice go away? No. But action created evidence, and evidence turned the volume down. I didn't conquer fear; I out-delivered it. Growth felt like stretching, not tearing, and the feeling of "not worthy" had fewer places to hide.

Takeaway: Fear of failure is normal. Let it ride shotgun, not steer. Shrink the problem to the next 48 hours, ship one real win, and stack receipts. Confidence follows the work, not the other way around.

We spent years waiting to be ready, always telling ourselves, "One day." The promotion taught us the trick is to start anyway: "day one."

Confidence follows the work, not the other way around. The day that lesson branded itself on us was the day we stopped outsourcing our agency.

During my offboarding at Keurig, I sat with a senior executive and unloaded a tidy monologue about everything broken in the business. They listened, patiently and silently, then asked one question: "What specifically did you do to help fix all those issues?"

That question hit me in the sternum, and I felt it move all the way into the pit of my stomach. I'd been tagged a "top 40" talent just two years into managing people, and I thought I knew it all. The truth? I wasn't powerless, and I hadn't done a single thing except complain. I had become what I dreaded most: a victim.

A victim of the system that was set up to make me fail, and of which I was powerless to change. With one well-placed, reflection-inducing question, I was set straight. It was always in my control. I must be the agent of change. It's a choice, but once accepted, the onus and responsibility are mine to lead with bravery and courage necessitated by leadership.

Yelena: That's when our house rule began: *If you're not changing it, you're choosing it.* We've carried that rule ever since. We now aim to be part of every solution, but more importantly, we refuse to be part of the problem.

Pro tip: If you are not part of the solution, you are the precipitate. For those of you not in science: The precipitate often gets discarded. Choose to stay in solution.

Pro tip #2: Failure's job is to teach; ours is to get up fast enough to learn.

When Your Data Becomes a Dead End

Sometimes failure doesn't come from sitting on the sidelines; it comes from being so sure you're right that you accidentally take yourself out of the game. I was reminded of this recently when speaking with a mentee who had just been laid off.

Their company had invested heavily in a shiny new herbal ingredient for a supplement line. It looked great on the marketing deck, except it was a nightmare to work with in tablet form. My mentee knew of a different, more functional format that would have solved the problem instantly. They had data to prove it. They presented passionately, repeatedly, and with absolute certainty.

Unbeknownst to them, though, the company's main mission was tied to that ingredient in that form. It was part of their brand story, their supply agreements, and their investor pitch. In other words, the ship wasn't turning.

Instead of pivoting their approach to find middle ground, my mentee doubled down, insisting the company was "wrong" and that their way was the only viable path. They didn't just present data; they fought the direction. They became the obstacle instead of the solution. Eventually, they became expendable.

The hard truth? Sometimes, business reality is bigger than your pet project, your data set, or even your truth. If your value to the company is only tied to being right, you'll have a hard time surviving when "right" and "strategic" don't match.

The better path, and what I encouraged them to do the next time, is to find a company whose mission and values align with their instincts, while also learning the soft skills to influence without alienating. If science doesn't fit, the business still fails, no matter how perfect the p-value.

Failing is scary. It stings. It's also the cheapest teacher you'll ever have. When a toddler wobbles and drops, no one says, "Stay down." We steady them and say "again." Adulthood complicates the stakes—mortgages, managers, pride—but the rule doesn't change. In the lab, we call it data: get up, adjust, and run the next trial.

For example, going through a divorce to learn a lesson on long-lasting relationships is costly and traumatic, to say the least. I speak from personal experience, long before Yelena and I built our life together. Yet, failure unlocks clarity, insights, resilience, and wonders that are priceless. Failure doesn't just toughen you; it re-shapes you. It strips away what doesn't matter and forces you to rebuild with what does. That's the hidden gift, if you choose to take it.

Sometimes, Yelena and I feel like our kids have it too easy. It's counter to every fiber in our bodies that's telling us to protect them from everything, even a fly buzzing around their heads. When we talk about what we can do better as parents, we often wonder out loud: *How can we help our kids get that "fire in the belly," the grit they'll need to succeed, without having to watch them fail?* We know they will fail—it's unavoidable. We know those failures will teach them more than any lecture we could give, shaping them into tougher, wiser versions of themselves. But the parents in us still want to shield them from it, to stand in the gap and catch them before they hit the ground like we've done countless times.

Once, if you recall, we literally did. On the shoulder of I-93, in a moment that felt like both the longest and shortest seconds of our lives, we caught our daughter as she entered the world. No hospital lights, no steady beeping monitors, just the cold "Raine" (her middle name), the sound of passing cars, and the undeniable instinct to hold on tight. That moment etched into us the desire to always be there, arms ready, no matter what was falling toward us.

Life doesn't work that way, though. We can't stop every stumble; we can't soften every blow. Maybe writing this book is our way of trying, by leaving behind a playbook they can turn to when life knocks them over. We know it won't stop the failures, but maybe it will help them see that failure isn't the end. It's the training ground.

Just like in the lab, the best formulas aren't the ones that worked perfectly the first time; they're the ones that broke, got messy, and were reworked until they were better than before. That's how resilience is made. That's how they'll be made.

It's taken us a long time to make peace with the idea that failure can be a friend. We've always tried to play it safe, even when our careers or personal lives seemed full of risk from the outside. The truth is that it takes a strong mental state to fail well. Letting go of pride and fear runs counter to instinct.

We understand why some people listen to those protective instincts and choose the comfortable course. Comfort comes with trade-offs, though. It can slow growth. It can leave space for envy and jealousy to creep in. It can rob you of the kind of resilience that only comes from getting knocked down and deciding to get back up, stronger.

Here's the move when your data becomes a dead end: Treat it like a detour sign, not a tombstone. Change the variable. Reframe the question.

José: I don't spend much time dunking and bashing on leaders who take big swings (like billionaires, entrepreneurs, and politicians). Risk has costs, and people who step onto the field take the hits. In a system that rewards initiative, choosing risk (or choosing comfort) both make sense. Just be honest about the trade-off.

The point I'm suggesting is that there are incentives for those who take risks in the face of failure. Risk is not for everyone, but then the rewards shouldn't be either.

Pro tip: Fail fast, get back up, and make new mistakes—don't remake old ones!

The Yelena Factor

I've lived this from the inside. I've been in conference rooms where scientists dig in so hard the conversation stops being about solutions. My question that unsticks the room is simple: "Are we protecting the data or solving the problem?" I've learned that influence beats insistence. My approach is simple: deliver the science in a concentrated, potent way, but blend it with enough business milk and sugar that people will actually drink it. I call it the "Espresso Shot of Truth."

Short Stories on Failure

An engineer, an analytical chemist, and a product developer walk into a Midwest bottling plant. Sounds like the start of a bad joke, right? Unfortunately, it was my reality.

We were chasing a mystery: a mid-calorie soda that wanted to audition for Old Faithful during pre-launch line trials. Every cap twist—a geyser. For days, we showed up like crime scene investigators, armed with clipboards, samples, and the feeling that something simple was hiding in plain sight. The culprit wasn't exotic chemistry; it was a dosing variance in a standard antifoam processing aid. With line time tight and the schedule compressed, we hadn't replicated enough pilots ahead of the trial. We corrected the dose, tightened the checks, re-ran it, and the geyser went quiet. No product reached the consumers.

Lesson: Skipping a proper test because it's convenient is the most expensive shortcut. Build the time in, or you'll spend it later, with interest.

Not long after that, I flew to the Chicago area—technically Barrington, but Chicago will do. This trip wasn't to solve a product mystery; it was for a possible career change.

I was under a management style that didn't match how I work best, and I wanted a fresh challenge. When an interview opened in the sports-drink division, I said yes, even though my heart wasn't quite in that particular role. Anyone who's been on the other side of the table can spot that. Midway through, the interviewer noted, fairly, that my energy didn't align with the job. Frustrating to hear? Absolutely. Accurate? Also, yes.

Back home, a senior vice president asked why I'd been exploring. Instead of bluffing, I said exactly where I wanted to contribute inside the company. To their credit, they listened. I moved into the role I actually wanted, and I spent the next two years doing some of my best work with a team that fit from Day One.

Lesson learned: Chasing an opportunity just to get away from something rarely works. But knowing exactly where you want to go? That can change everything.

Honestly, that lesson doesn't just apply to career moves; it applies to life. Let's be real: Life often feels more like a juggling act than a casual game of catch. Most people are doing it while running uphill, in the rain, with one shoe untied. You're trying to heal from the past, shake off a rough childhood, excel at school, perform at work, be a present parent, hydrate, return texts, answer emails, keep up with calls, and somehow look like you've got it all together on social media—all while remembering to buy milk.

It's a lot.

While you can't control all the chaos, you can decide how you approach it. For us, a few small steps have made all the difference:

- **Small wins:** Stack enough of them and they start to feel like momentum.

- **Iterate:** Nothing has to be perfect the first time—keep adjusting.

- **Reflect:** Pause often enough to understand what's working and what's not.

- **Have fun in as many things as possible:** Even the messy parts.

Or, to borrow from Gary Vee: "Fall in love with the journey." The wins feel sweeter, the setbacks will sting less, and you'll start to realize you don't have to be perfect; you just have to keep moving forward.

If you're looking for perfect, we're not your people. If you're looking for repeatable moves that work under real constraints, pull up a chair.

Why take advice from us? Because we've lived it together. We've built careers in high-pressure, high-expectation environments while raising four kids, side by side. We've juggled product launches, cross-country flights, and midnight crib runs—one of us shushing, the other rocking—like a relay team that knows the handoff in the dark. We've stood shoulder to shoulder in kitchens, airports, and labs, picking each other up when projects blew up (sometimes literally) or when plans veered off course. We're not perfect, but we've learned to play to each other's strengths, laugh through chaos, and keep moving.

We're not offering a "10-step program to a perfect life." That doesn't exist. We're offering the steps that kept us moving when things got chaotic. Some days, people need constant encouragement; we did, too. On those days, we borrow courage from each other and ship the next small win.

What We Won't Do:

- Sell you a hustle as a personality

- Pretend trade-offs don't exist

- Talk like unicorns when we're just consistent

If any of this earns your trust, good. If not, we'll still leave you with tools you can use tomorrow. That's why we wrote this: not to be right about everything, but to be useful to someone building under real-world pressure.

Take what helps, ignore the rest, and start where you are. Day One is the only day that changes anything.

From Obstacle to Asset

If you're a scientist in business, here's the uncomfortable truth: Your job isn't just to be right. Your job is to be valuable.

Valuable means your expertise helps the business move forward. Sometimes that means adapting your recommendation to fit cost realities. Sometimes it means offering a "good enough" option to keep momentum while you keep advocating for the optimal one. And other times it means letting go of your favorite solution because it's a mismatch for the company's mission. Here are some steps to follow in case you need a guide:

Mini-Framework: The Four Steps from Stubborn to Strategic

1. **Pause:** Before reacting, ask: "What's the bigger picture here?"

2. **Understand:** Learn the market, consumer, regulatory, and cost constraints driving the decision.

3. **Reframe:** Position your data as an ally to the company's direction, not an obstacle.

4. **Influence:** Offer solutions that balance scientific integrity

with business momentum.

Finding Your Fit

Sometimes, no matter how well you influence, you're just in the wrong place. If the company's mission will never align with your science, it's okay—and healthy, even—to find a place where your voice is valued.

Wherever you land, remember that business reality is part of the equation. Influence travels farther than stubbornness. Finally, science, when it works with the business instead of against it, changes the world.

Earlier in the book, we introduced our Recipe for a High-Performing Team:

- **1 part "positive manager(s)"**

- **2 parts "no egos"**

- **3 parts "curiosity-filled mindset"**

- **Garnish with a sense of humor**

It's a simple, powerful formula, but even the best recipe can be ruined if one ingredient dominates the dish. In corporate science, one of the biggest "overpowering flavors" is scientific stubbornness.

How the Recipes Connect

- **Positive manager(s):** The best leaders help bridge the science-business gap, as long as it's a two-way street. Even the most positive manager can't help if you're actively resisting alignment.

- **No egos:** This isn't just about avoiding arrogance. It's also about staying open to other perspectives, even when you believe your data is flawless. If you can't let go of your version of the "truth" long enough to listen to business realities, your ego is quietly taking over the dish.

- **Curiosity-filled mindset:** True curiosity doesn't just encourage you to explore the science. It pushes you to understand why the business might be heading in a different direction. What market forces are driving this? What supply chain constraints exist? What do consumers actually care about here? Curiosity keeps you from falling into the "I'm right, they're wrong" trap.

- **Sense of humor:** Humor lightens the mood, but it also builds resilience. If you can laugh about the fact that your perfect, elegant, tablet-friendly herbal form is getting replaced by something clunky because it "tells a better marketing story," you're halfway to surviving another day in corporate science.

From Kitchen to Conference Room

If your personal recipe has too much "stubbornness" in it, you risk turning into the overly spicy jalapeño of the team: memorable, but not always in a good way.

That's where the *Barbosa Stubbornness Quotient* comes in. Just like balancing flavors in a lab, you need to balance passion with adaptability.

The ideal scientist in business isn't the one who's always right; they're the one who makes the whole dish work.

The Barbosa Stubbornness Quotient (BSQ™)

Think of this as a companion scale to the Barbosa Resiliency Quotient:

BSQ Level	Symptom	Corporate Translation	Career Outcome
1	Calmly presents data, open to alternatives	"Team player"	Trusted voice
3	Subtly digs in, re-emails charts	"Persistent"	Mixed influence
5	Brings peer-reviewed papers to every meeting	"Difficult"	Fewer invites
7	Publicly questions strategy in cross-functional calls	"Not a fit"	Exit interview
10	Storms out, muttering about how their tablet-friendly herbal form will save humanity	"Security?"	Out the door

Barbosa Stubbornness Quotient (BSQ): How corporate science interprets scientific persistence

The sweet spot? BSQ 2–4: passionate enough to care, but flexible enough to stay employed. Let me share something I've come to realize: The moments you'll be proudest of in your career are rarely the times you muscled through a miserable environment just to prove you could, like taking organic chemistry three times just to pass it. They're the moments where you showed up fully, bringing curiosity, clarity, humor, and yes, kindness.

So, now that we've talked about failure, flexibility, and finding that BSQ sweet spot... let's talk about the bigger prize: being your best self, at work, at home, and in life.

Chapter Twelve

Seasoning to Taste: Becoming Your Best Self

José: From billionaires and burnout, let's pivot to something far more energizing: your best self.

What does it take to be your best self at work, at home, and in life? Many experts will say psychological safety, which Yelena and I could not agree with more. Those who've visited malnourished communities will remind you that health is also paramount. Just think about it: When you're sick and working, you don't feel "whole," right? Finally, I'll add a third ingredient: self-worth.

Know this: Every single person reading this (and even those who aren't) has incredible worth—to others, to themselves, and yes, to humanity. Our collective consciousness benefits from every single one of you striving to achieve your best self. We hope every one of you has someone in your life who reminds you of that. If not, reread the last sentence and consider this our sincere thank you for being *you*.

Yelena: I'll double down on that. Worth isn't determined by a job title, a specific salary, or the number of LinkedIn endorsements you can obtain. It's knowing that you matter, even when you are not "producing." We've both had times where the work itself was exhausting, but feeling valued, or at least seen, kept us going. When that wasn't happening, we had to create it for ourselves and for our teams. We try to ask empathetic questions, which can feel challenging at times. We don't always have finesse. We've also learned to become cheerleaders. Saying, "You're doing great" after someone has made a mistake is not easy, but when you try it, you see the effects immediately. They might even realize they've become their own worst critic. While they might have criticized themselves for that same mistake, others are celebrating their effort. That's a good place to be.

Unlocking Your Best Self at Work

José: I've always aspired to create a kind of workplace where people can operate with clarity, curiosity, and passion. It's easier said than done, but it's essential for career success.

Operating with clarity is a secret of many successful individuals. Responsibility for obtaining that clarity lies with both the leader and the employee. My toughest moments came when I didn't have the clarity I needed or didn't ask the right questions. As a leader, it came when the team delivered on something I didn't need but thought I had requested. Pride can be expensive. Unless you're pitching to investors, ask the "silly" question. In science, and in most of life, it's usually not silly at all.

Yelena: I'll add that clarity is a two-way street. I've been guilty of thinking, *If I just work harder, they'll notice.* Spoiler: They don't. People are too busy putting out their own fires. You have to articulate your needs and your confusion. Don't assume. Don't remain silent. Silence isn't a strategy; it's invisibility.

José: One of my superpowers? Kindness. I talk to people. I listen with intent. I offer a smile—after all, why not? It's free! You'd be amazed at how far a sincere and authentic smile can go.

If I ever studied human psychology, I'd want to understand which emotions drive behavior. Fear would probably top the list. My hypothesis is that it dominates only because we're still maturing as a species. My hunch? As we evolve, kindness will overtake fear as the more powerful driver.

Yes, I know I'm saying that while the world is more divided than it's been in decades, but I've seen how kindness and dialogue work, up close. Fear has bona fide enemies.

The Kindness Shortcut

A typical product launch requires multiple functions doing their part in sequence. Dates are set but often treated as guidelines rather than stone-carved commandments.

I always made it a point to know each person in the process. Not just their job titles, but them—their quirks, coffee orders, favorite holidays (Halloween is the unofficial preferred one). So, when it was their turn to move my formula forward, I didn't send an email reminder. I walked over. I said hi. We talked, and before I left, they'd done their part.

What normally took two weeks, I could finish in a day. Not because I "worked the system," but because we were friends. You can fake enthusiasm, but you can't fake friendship, not for long anyway. Authenticity is a foundational ingredient in life.

Yelena: That's not just a Hallmark sentiment. In my experience, kindness makes you memorable in the best way. It opens doors that "authority" alone can't. I've had project bottlenecks magically disappear because someone genuinely wanted to help me, not because they had to (thank you! You know who you are!).

When Business Is Good

José: The Lorax famously said, "Business is business and business must grow, regardless of crummies in tummies, you know." Yet business doesn't have to be bad.

When people bring their best selves to work, mountains turn into molehills. Work can create belonging, pride, and shared purpose. The best teams I've been on genuinely enjoyed being together.

Maybe artificial intelligence (AI) will one day allow us to sustain ourselves without an eight-hour workday. Maybe then I'll stop praising the corporate world, but for now, I believe business can be a force for good if we lead with empathy and kindness.

Kindness ≠ Weakness

In corporate culture, "too nice" can be code for weak. Sometimes, kindness gets tangled up with humility, causing you to avoid self-promotion and let others speak for your accomplishments. That's noble, but in the corporate world, it means your wins go unrecognized.

Growing up in a Hispanic community, I learned that others should speak of your accomplishments, or you risk arrogance. I learned to respect my elders, which made decisions such as laying off a mentor particularly challenging and even painful.

Over time, I learned to separate corporate norms from life values. Yet I never abandoned kindness as strength. I still led with urgency. I still tackled conflict head-on. I still held courageous conversations. Kindness and strength are not mutually exclusive.

Yelena: People like helping people who treat them as people, not as cogs. Friendship in the workplace isn't a luxury; it's a hidden efficiency engine. And let's be honest: Being kind doesn't mean you avoid conflict. I've had some of my toughest, most direct conversations with people I deeply respect. Respect and kindness make conflict productive instead of toxic.

Pro tip: Be kind.

Bring Value or Bring Change

If you can't find it in you to be useful at work, and if you're not happy, productive, or contributing, maybe this career isn't for you. That's okay. Change careers. Just because Yelena and I love it doesn't mean it's for everyone.

If you do belong where you are, ask yourself: *What's stopping me from being a superstar?* Do you need a role change? A new skill set? A fresh perspective?

Whatever it is, bring value and you'll be valued. The people who feel guilty taking a paycheck usually feel that way because they're all-in; they know they're contributing, creating, and delivering, and that's fun. You're getting paid to have fun. Instant guilt, but bullseye reward.

If you're struggling to feel valued, remember, it may not be you. Sometimes it's the environment. When in doubt, take my parents' advice: *Be the best at what you do. Always.*

Pro tip: Be useful.

Yelena's Approach to Growth

When I first started at PepsiCo, "career growth" wasn't even a phrase I used. I was a young scientist over the moon to have a real job at a company my family recognized from across oceans. They didn't fully understand what a food scientist did, but they knew the brands. That cheering section was enough.

My focus was simple: do the work, learn fast, belong. Growth didn't mean climbing; it meant getting better, and showing up with curiosity and energy.

Over time, I learned the rhythms of corporate life. Hard work matters, but intention moves you. Growth asks for stretch assignments—for leading before you're given the title, for thinking ahead in honest timeframes: What do I want to be doing in 2–3 years? What might 10 years ask of me?

Each role taught me science and people. I learned from managers I admired, and from ones who showed me who I didn't want to become. Growth felt organic until a conversation nudged it into focus.

A senior leader asked me, "Do you want to be a CEO one day?" I paused and said no. At that time, "CEO" meant constant travel and a life far from a future family. Later, I learned that "CEO" is a shape, not just a plane ticket. It can be a small regional company, a tight-knit team, or your own venture. That reframed everything.

Today, growth for me isn't a ladder; it's a landscape. Titles are one feature, not the destination. I want both a career and a life I'm proud of. Real impact, relationships that matter, and challenges that keep me learning. That's not a lot to ask for, right? Growth is multidimensional now: professional, leader, parent, person.

If there's one lesson I'm holding close to my heart, it's this: Growth isn't a place you arrive at; it's a practice you keep. Stay curious. Stay connected. Keep pushing forward from where you are. Thankfully, growth is endless.

Part 4: Dessert & Digestifs; Reflections From The Trenches

Chapter Thirteen

The Business Banquette: The Economics of Food Science in Business

José: During the later stages of my career, I discovered how to put product launches into overdrive. In the supplement space, launching products could take up to 18 months or even longer! With smaller, nimble competitors popping up daily, this quickly became a significant competitive disadvantage, an Achilles' heel. Small companies would find overnight success and sell for millions in a short amount of time—just look at Smartypants and Olly, to name two. To compete with this trend, our team went on to launch over 30 products in 22 months, a feat that only a few agile competitors could match or exceed (I'm leaving it open to that possibility, however unlikely that is)!

You might be wondering: How did we do it? And was it enough to grow the business? (Heads up: I don't know the answer to this last one.)

How did we do it? Well, like the movie *The Martian*, we kept solving problems until we had solved enough to take off like a rocket ship. Fueled by clear direction from the top of the hierarchy and the propulsion of great team dynamics, we reached liftoff. With a true understanding of how people and relationships work in business, we kept molehills from becoming mountains and greased the right wheels to move forward with external partners. Internally, we mapped out who did what and how it all fit into the bigger priorities. The outcome? A novel approach to aligning R&D with the commercial side that accelerated success. One can argue that's only part of the equation—and you'd be right—but celebrating those wins created the momentum to keep building. Over time, that compounding effect might just build a business that outlasts us all.

If you're joining the Consumer Packaged Goods (CPG) space today, know that these soft skills matter more than ever. AI might someday manage a stage-gate system or even pitch its own product. What it still can't do is build trust, navigate nuance, or lead people in real time. That's your advantage—don't forget it.

Consumers are reshaping the food landscape daily. Just look at how players like Whole Foods, Trader Joe's, and Sprouts have grown into major forces over the past 30 years. What about the next frontiers? Personalized nutrition delivered to your door via drones or nutrition that allows astronauts to reach Mars. If Elon Musk gets his way, food scientists may soon be developing tastier and more practical foods for Mars. I, for one, can already taste the freeze-dried *mofongo*.

The Mental Trap of the Golden Goose

Are you ready to talk about something a little taboo? Let's address the elephant in the breakroom: salary. Yep, I'm talking about mine. Not to flex, not to humblebrag, but to explore how chasing numbers can sometimes dull the edge of purpose.

Food scientists are, in our opinion, well compensated, especially as their expertise deepens. Flavorists, those who engineer taste itself, are and have always been in particularly high demand. It's a niche role requiring deep chemistry knowledge—specifically organic chemistry—and that specialization drives real value and high compensation. It is a very specialized field and critical to the "taste is King" mantra. Now, how about a product developer or a formulator? A quick search shows that many food scientists in 2025 averaged between $75K and $90K. That may not make headlines, but it's a solid, fulfilling income for a career that feeds (literally and figuratively) millions. That's a great place to start or build a career, and yes, it can grow.

Now, for the personal disclosure that will probably be the topic of a few social dinners. Although it's risky to say aloud in today's soundbite culture, here it is: At certain points in my career, I earned what most would consider a life-changing salary.

I know that statement might sound vague or provocative. But I'm not sharing it to brag. I'm sharing it to tell the truth about something people don't talk about enough: the hidden cost of high compensation—or, more broadly, the hidden cost of any compensation.

There's a concept called golden handcuffs. It's the idea that the more you earn, the harder it becomes to walk away. The paycheck grows, but so does the inertia. At a certain point, your income stops being a tool and starts being a trap. You begin to confuse fear with prudence. I've lived it.

The bonus becomes addictive. The title becomes your identity. You tell yourself you're staying because it's smart, but really, you've built a life that is too expensive to rethink. Your career starts serving your lifestyle, instead of your purpose.

Now, don't get me wrong: I'm not saying don't chase success. Chase it. Break records. Get that promotion. Just do it with your eyes wide open. Because the same ambition that fuels your rise can quietly become the leash that holds you still.

When the moment comes to bet on yourself, the ability to walk away isn't weakness; it's wisdom.

And here's the truth: Golden handcuffs aren't just for executives. They can tighten at any salary level. Whether you're making $60K, $160K, or something with two commas in it, if your lifestyle, identity, and peace of mind are tethered to your paycheck, then yes, you've got handcuffs on too.

If you're early in your career and wondering how this applies to you, remember that money is a tool. It doesn't care about your stress. It won't cry when you burn out. Learn to use it, without letting it use you.

Money is not the enemy. But it's also not the goal. Growth is the goal. The salary will come if you're valuable, curious, happy, and open to learning. Just don't lose yourself in the title. Because if you're not careful, you stop asking, "Am I fulfilled?" and start asking, "What happens to my stock options if I leave?"

Pro tip: Money is emotionless, so manage your relationship with it like a business, not a romance. Earn it. Enjoy it. But never let it own you.

Some readers will want to know the exact number. That's understandable. But I've learned that sometimes, the number overshadows the story. What matters isn't what I made—it's what it made me.

(Shared to inform and empower, not to impress. Context matters.)

Chapter Fourteen

Kitchen Dynamics: People, Power, and Pipettes

The Good, The Bad, and The Micromanaged

Who really has influence over your salary, your happiness at work, and your opportunities? Your manager, of course. Unless you start your own business, you're going to have one, and like it or not, they will shape a large part of your career.

Between Yelena and me, we've had over 30 managers. We've seen the good, the bad, and the truly micromanaged. Sure, you can find plenty of memes that capture these archetypes in a single punchline, but nothing beats real-life stories.

The Good

Have you ever had a manager push you in exactly the right way, even if their suggestion annoyed you at first? That was me when my manager told me, "You need a master's degree in food science if you want to keep succeeding in corporate America." Those were not the words I wanted to hear, but they were golden advice.

Yes, I got that master's degree. It only took 10 years and constant "encouragement" from Yelena (let's stick with that word since she's co-writing this book. Keep that between us, okay?).

My master's degree from Rutgers University was anything but traditional. I wasn't a campus regular; in fact, I could count on two hands the number of times I set foot on Rutgers ground. Most of my "classroom" time took place from a conference room in PepsiCo's Valhalla R&D campus, where the class would log into live lectures over video at a time when remote learning technology was still clunky, expensive, and far from mainstream. It was a strange marriage of corporate grind and academic pursuit, one minute tasting flavor prototypes in the lab, the next watching a nutrition lecture through a surprisingly clear and vibrant screen.

By the time my final presentation rolled around, I knew it had to be more than a slide deck. My thesis focused on muscle hypertrophy and the role of proteins in combating sarcopenia, the muscle loss that comes with aging. So, I decided to make it... delicious. With Yelena as my secret weapon, we created a menu that translated my research into something tangible, edible, and memorable. We cooked and plated foods rich in the protein profiles I'd been studying, giving the professors not just data, but a sensory connection to the science.

I essentially had fun channeling my passion, food science, into my deliverable.

The presentation landed better than I could have imagined. The professors leaned in and asked questions. They ate. They smiled. By the end, their feedback made it clear: This was more than a thesis defense; it was a masterclass in making science come alive.

I'll be honest: Up until that moment, I hadn't been the star student. My grades were fine, but not stellar. This presentation, though, was my redemption arc. It proved to me that when you merge technical knowledge with creativity, and when you feed people the science instead of just telling them, you don't just earn a degree. You earn connection, respect, and the chance to inspire. For that, Rutgers and their stellar Food Science program will always have my gratitude.

In hindsight, my manager's advice was exactly what I needed, even if I didn't want to hear it at the time. Sometimes the best leadership doesn't just point you toward a goal; it gives you the push you need to take the first step.

The best managers share traits that will sound familiar if you remember the "recipe" I shared earlier in the book: no egos, strong skills, and the ability to have fun without crossing professional boundaries. They trust people who trust themselves. They give you space to stretch your abilities, and when you stumble (because you will), they offer cover instead of criticism.

That's how I became the first technician in my era to launch a product: KAS Pink in Mexico. My manager believed I could do it, took a risk, handed me the project, and then guided my overly caffeinated enthusiasm with just enough realism and a smile.

Yelena: My best managers made me feel like I was the one steering the ship, even though they were quietly handing me the map. They didn't hover. They didn't second-guess every decision. When things went wrong, they stood next to me, not six feet away, pointing at me. That's how you earn loyalty that lasts long after the org chart changes.

The Bad

I could fill an entire chapter on the next two archetypes, but I'm not one to wallow in negativity. So, I'll keep it brief and offer a perspective that allows us all to learn what not to do, especially if we're exhibiting these behaviors as we read.

One of the worst types of managers is the "Do-Nothing Boss." They're paralyzed by fear—fear of taking risks, fear of making mistakes, fear of deciding at all. Their inaction drains a team faster than a slow leak in a soda fountain. High performers suffocate under their watch because nothing moves forward.

Retired Air Force Major General Perry M. Smith called them "wimpy bosses" in his book *Rules & Tools for Leaders*. I call them career quicksand.

Yelena: My personal favorite is when they say, "We're still thinking about it"...for six months. At that point, it's not thinking—it's avoiding. The longer they avoid, the more your project and career rot on the vine.

The Micromanaged

This word really does say it all. Micromanagers hoard decision-making like a dragon hoards gold. Every success belongs to them; every misstep belongs to you. They are convinced you could never possibly find your way without their guiding hand.

Their egos are as fragile as a toddler's feelings when you cut their sandwich into rectangles instead of triangles. We have four kids, so this analogy hits home. General Smith refers to them as "Big Ego Bosses," and his description could make even the highest scorers of the Barbosa Resiliency Quotient (BRQ) scale sigh in despair.

Yet, micromanagement is rarely a product of ego alone; it often stems from fear. Fear of the project failing. Fear of how it reflects on them. Fear of losing control, which ties into power. I get it. I've been there. The irony is that the tighter you grip the wheel, the more you risk steering the car into a ditch.

Most of us who have ever micromanaged didn't wake up deciding to do it. We were trying to protect the work, our reputation, or our team from mistakes, but we didn't realize the cost.

Some micromanagers, often without even realizing it, seem to want more than just compliance. What they're really after is something closer to capitulation. Not the kind you read about in history books with flags of surrender, but the quiet, unquestioning acceptance of "their way is the way."

It's rarely a conscious demand, but it can manifest in subtle ways: constant overrides of your decisions, "corrections" that undo perfectly good work, or the feeling that every initiative must trace back to its approval to have any value. Over time, the unspoken expectation becomes clear: They want your work, your approach, and sometimes, even your thinking to bend entirely to theirs.

The irony is that this desire for total alignment often works against them. Capitulation might bring short-term compliance, but it kills the spark of creativity, problem-solving, and ownership that great work requires. In healthy leadership, influence flows in both directions. The best managers don't seek unconditional surrender; they seek partnership.

So, if you find yourself with a manager whose style leans toward this quiet demand for capitulation, try to engage them in dialogue, show them the value of shared ownership, and protect your ability to think independently. You might just help them see there's more power in collaboration than in conquest.

I recently spoke to a former colleague, a superstar when they worked for me, with credentials among the best in the industry and experience in both technical and commercial roles. Yet they told me they could feel it: Their insecure manager was preparing to throw them under the bus, and not for the first time. This manager had a track record of sacrificing others just to look good. My former colleague is now planning an exit.

Sadly, I could relate all too well. The corporate world still teems with managers like these, and surviving them takes a special mix of self-awareness, resilience, and the ability to recognize when to move on.

Yelena: Micromanagers always say they "just want to help." That's like a cat saying it "just wants to play" with a mouse. Either way, it's not ending well for the mouse.

Early in one role, a manager asked for weekly updates, which I found mildly annoying but doable. Then the cadence shifted to daily logs, line by line. At that point, a restroom break felt... loggable. There wasn't space to propose a new system; the style was the style. So, I did the only thing I controlled: the work, my attitude, and eventually, my environment. I transferred to a different team (which is one reason big companies can be a gift). Not everyone has that option in a smaller organization, and I don't take it for granted.

Lesson: Sometimes you can't change the manager or the mechanism. You can still change your lane. That's not quitting; it's choosing.

In another instance, I had to navigate another rocky relationship, and it felt every bit as bruising as being in the ring with Rocky himself. Navigating managers in the corporate hierarchy is no easy task, and no one really teaches you how to do it. After a few career jumps, I found myself reporting to a manager who inadvertently, and, let's be honest, quite painfully, helped shape the kind of leader and manager I am today. Our relationship was tough. We couldn't see eye to eye, and our communication styles clashed in a way that didn't just block productivity but also fueled tension. As the months turned into years, the tension only grew. I remember feeling anxious before our weekly one-on-ones, never sure if I was walking into a neutral check-in, a slightly positive conversation, or more often than not, an intense exchange that left me rattled.

Like many of us in the corporate world, I'd done my share of personality assessments—Myers-Briggs, StrengthsFinder, you name it. I'm the type who likes to ease in with a little small talk, then get to business. Over time, I realized that my manager functioned best with quick, bullet-point updates—no fluff, no detours. If I stuck to a tight agenda, we could get through the meeting with minimal friction. That approach worked for a while. Eventually, whether by luck or strategic reshuffling, I was reassigned, and things got easier.

Even with a more compatible manager, I learned an important truth: Every new reporting relationship requires a period of recalibration. You have to learn how the other person thinks, communicates, and makes decisions. There's no shortcut. If you want your projects and your career to thrive, adapting your style just enough to meet them halfway is part of the job.

José: I've been there, which is why I tell my teams: no surprises. You'll always know where I stand and what good looks like, so one-on-ones aren't guessing games.

We can't rewire every insecure manager, so let's focus on what we control.

This is where your BRQ matters. If you're running high on resilience, you may have the appetite for bigger moves, like shifting functions, relocating, or even starting something of your own. They're scary, yes, and they're also some of the fastest ways to reclaim freedom.

If your BRQ is still growing, this means you're not quite ready for those big swings, but don't worry. You still have powerful tools in your corner: deliver exceptional work, check in often with your manager so there are fewer "surprise" moments, and pivot quickly when priorities shift. Think of it as corporate judo, using the momentum of the situation to your advantage instead of fighting it head-on. (Ironically, my dad signed me up for endless judo lessons as a kid, and I still can't break a napkin, but I can roll with corporate throws like a black belt.)

No matter your BRQ score, the key is staying in motion. The real trap is stagnation, not your manager. Remember: Every frustrating manager you survive is just another rep in the gym of your corporate resilience.

If you recognize yourself here, you're not alone. Every great leader I know has had to unlearn micromanaging tendencies at some point. The ones who did found not only more capable teams, but also a lot more sleep at night (although they also had an additional variable of taking magnesium, so it could have been that too).

I'm sure it's crossed your mind to ask, "What is it we're supposed to do, both as leaders and individual contributors?" Well, as a leader, if you've ever worried about letting go, try this: pick one project or one decision and let your team run with it, without stepping in. Set a clear outcome, give them the resources, and commit to being hands-off unless they ask for your help. You may be surprised by the results. Now, allow me to suggest the following behaviors for the individual contributor—which applies to leaders too—that you should avoid at work and in life.

Victimhood: The Most Comfortable Pair of Slippers You'll Ever Regret Wearing

Let's talk about a mindset I've come to see as my personal kryptonite: the victim mentality. You know it. You've heard it. You've maybe even worn it like a bathrobe on a lazy Sunday, muttering, "This always happens to me." "They never listen to my ideas." "The system is rigged." (Okay, maybe sometimes it is rigged, but you still must play the game.)

Victimhood is wildly comforting at first. It requires no action, no accountability, and no effort, just a big ol' beanbag of blame and a pint of "not my fault" ice cream (which is the worst ice cream, by the way). Much like my kids' "off the chart" energy levels after eating said ice cream, it will come back to bite you in unexpected ways. It saps your energy, your confidence, and, most dangerously, your belief that you can change anything.

Now, before anyone starts composing an angry email with bullet points and footnotes, let me be clear: There's a big difference between acknowledging real hardship and choosing to live in a state of permanent helplessness. Life isn't fair. Work isn't fair. Sometimes, you do everything right and still lose. I'm not denying reality; I'm inviting you to rebel against it.

Victimhood is passive. Growth is active. You don't need to be a superhero to shift your mindset. You just need to take one small, uncomfortable, possibly sweaty step forward and ask yourself, "What can I control?" Spoiler alert: It's probably more than you think.

So, if you find yourself slipping into the role of "the overlooked, the unappreciated, or the chronically coffee-deprived," take a breath. Take a walk. Take accountability. You've got this. If you're not sure where to start, I'll give you a hint: It's not with them. It's with you. Instead of taking people down, lift them up. If you happen to feel and think like me, then let me hit you with some knowledge forged in experience: Bringing people down eats at your soul and never settles well in your stomach. None of us wants to use the word "regret" in life, and hurting someone's chances at success or someone's second chance brings you to the doorstep of regret. Don't do it!

Some of you have made the important decision to progress into management, whether it's the management of projects or even the management of people. Take a moment to congratulate yourself, and at the same time, brace yourself. If you've ever wondered how you can transform into a master leader, let us give you some food for thought. The following stories may help you digest what it takes to succeed on the leadership journey.

From Pipettes to Performance Reviews

Yelena: With each career move and every new company, I became more confident, more knowledgeable, and better at navigating the corporate world. As I worked my way up and stepped into management, I quickly learned that managing up and managing down are two entirely different skill sets. In my first role as a manager, I led a team that included both brand-new scientists and seasoned pros who could practically develop a beverage with their eyes closed. It was an exciting time, with equal parts learning and doing, as I figured out what it really meant to lead. From goal setting and midyear reviews to leading cross-functional teams and, eventually, navigating underperformance and even termination, I got a crash course in the full spectrum of people management. As anyone who has experienced it will tell you, your first termination is always the hardest.

The early months as a newly minted manager were a blur of trial and error. I was still managing my own projects while also learning how to prioritize strategically for the team, advocate upward to senior leadership, and build cross-functional relationships across R&D. Thankfully, I had a manager who was also a mentor, someone who helped me shape a team that not only earned respect from our peers but also consistently knocked projects out of the park.

What that stage in my career taught me more than anything else was that communication is everything. The good, the bad, and the awkward—it all comes down to how you say it and how clearly information flows in both directions. I had to fine-tune how I spoke to company leaders, striking the right balance between optimism and transparency, while also guiding my team to think bigger, prioritize smarter, and elevate their work. I found myself wearing my "business lens" more than I ever imagined, and to be honest, I loved it.

At a certain level in product development, your career path forks. One route takes you deeper as an individual contributor: a principal scientist, a subject matter expert, and the person who owns the bench and creates the magic. The other path leads to management. It's not for everyone, but for those who choose it, a few traits make all the difference: clear communication, expert time management, the ability to delegate effectively, and a healthy dose of self-confidence (without the ego). Of course, you also need to know the business deeply. That kind of knowledge isn't handed to you; it comes from experience, curiosity, and a genuine desire to understand how your company works, not just within R&D but across the board. The more invested and authentic you are, the easier it is to soak it all in and lead with purpose.

José: For us, a healthy work-life blend feels like a perfectly braided challah, each strand distinct but woven into something whole. On our best days, we can drop off the kids at 9 a.m., work for three hours, grab groceries, work for two more, pick up the kids, work again, tuck them in, sneak in another couple hours, then unwind together reflecting on our day. Yes, it's exhausting just thinking about it, but it's also the sweet spot where work and life don't compete for oxygen—they fuel each other.

While this blend works for us, it's taken years of trial, error, and spilled coffee to get here. That's why managing people, with all of their unique blends of work and life, is no small feat. We've been fortunate to lead incredibly professional teams throughout our careers, even as junior managers. Discovering how to lead, manage, and mentor isn't something you master overnight.

First, you have to learn to delegate. Then, you have to remember that your team members are not your kids; they're professionals, and they deserve to be treated as such. You must learn to give feedback in ways that inspire growth rather than fear. Most importantly, you have to develop the skill of dialogue, true two-way communication, rather than relying solely on one-way speeches disguised as "updates."

Ownership of the problem amongst those who lead is such a powerful motivator. Let me describe an example of when I inherited an "impossible" project: close out hundreds of old projects in record time. I sat my team of five down and simply asked, "How do we do this?" They designed the plan, set the rules, and even built in humor, like who would wake up anyone who nodded off mid-task. We agreed that vacation time during this sprint would be covered by others on the team, and that every member would have opportunities to "own the win" at different points in the project.

The result? We archived every single one of those projects and hit all of our quarterly goals for the day job. More importantly, we built trust, laughed through the grind, and proved to ourselves (and the business) that we could accomplish the improbable without sacrificing our sanity.

Delegation is one of the hardest skills for new managers to learn because the very thing that got you promoted to manager—doing your own job exceptionally well—can also hold you back. It's tempting to think, *It's faster if I just do it myself,* but that habit drags you back into the weeds. Our constant reminder to new managers: Get out of the weeds. My personal mantra for managers I lead: Get. Out. Of. The. Weeds. Did I say that already?

People show up wanting to matter. Give them a clear target, remove the blockers, and trust them with the work. They're not children; they're professionals. Lead with clarity and cover, not control.

Pro tip: If you're micromanaging or doing everything yourself because you think it's faster, you're not managing—you're just hoarding the work.

Yelena: I still consider myself early in my management journey, and I'm sure José could chime in with a few thoughts of his own on this topic. If there is one piece of advice I'd offer to anyone stepping into leadership, it's this: never speak poorly about your team in a cross-functional setting, especially in front of senior leaders. No matter how tempting it may feel in the moment, whether out of frustration, pressure, or a desire to explain a setback, there's no version of that story that ends well. It doesn't benefit you, it doesn't serve your team, and it doesn't reflect well on your leadership.

There's no such thing as a perfect team or a perfect leader, and we all know that, but choosing to highlight your team's strengths, resilience, and progress, especially when the spotlight is on, sets the tone for how others perceive your team and how your team perceives you. That doesn't mean ignoring challenges. It just means understanding when and how to talk about them. Sometimes, that means having honest conversations behind closed doors with your own manager, as you work to understand each individual's strengths, growth areas, and where they fit within the bigger picture. Other times, those discussions happen during team calibrations or performance reviews, where the full context can be considered and support plans put into place.

There's a time and a place for everything, but one thing that always holds true: be your team's biggest advocate. Show up with humility, honesty, and the desire to see each person shine, not because it makes you look good, but because it's the right thing to do.

José: The only thing I'll add is that in those meetings, you also obtain feedback about your employee or team, and it is critical that you bring that feedback to the employee in a constructive manner—not straight and cold, not soft and ineffective, just real. Easier said than done, I agree. Yet giving and receiving feedback is a gift. Wait, giving feedback is a gift? Really? Then why does it feel like passing a kidney stone? Too much good feedback and employees don't change any negative behaviors. Give too much negative feedback and you've demotivated them for the next few months. If you have someone on a performance improvement plan (PIP), then it becomes exponentially harder, as you must balance encouragement with direct and clear communication on where and how to improve. It's like trying to weigh a liquid using a pipette—one extra drop can ruin the entire mixture. You must get it exactly right, or the whole thing needs to be discarded. So, what makes for the best conversation? What helps you give and receive feedback? Allow me to extrapolate on a subject many of us think we know...

The Power of Dialogue

Dialogue is one of the most powerful skills, not just for food science and corporate, but for the entire human race (in my opinion). Learning to engage in dialogue means learning to listen, ask the right questions, and be open to other perspectives. You must put your ego away—there's that ego thing again—and converse in peace. This is not easy.

One skill has quietly shaped the best moments of my career, and it's not talking, debating, or presenting. It's dialogue.

When I found coworkers, employees, or managers who understood the meaning and the etiquette of true dialogue, the workplace transformed. The atmosphere shifted from transactional to collaborative, from compliance to creation.

I once had an employee who mastered this. They would listen, *really* listen, to my vision; then they'd process it, reflect, and come back with thoughtful counterarguments, backed by data and genuine curiosity. They didn't aim to win the conversation. Their goal was to build something better. And they did. We moved mountains together.

It helped that my manager was cut from the same cloth. The trust that flowed up and down the chain created a momentum that made growth and innovation feel inevitable. This wasn't just a good team; it was a great moment.

Did it last? Of course not. Business changes faster than you can update a slide deck. People move, priorities shift, and org charts collapse and reform like corporate Jenga. You parry with your best pivoting skills, and you carry on, sometimes for years.

However, unlike my philosophy, which is purposefully endless, even the best seasons of collaboration come to a close. Still, if you ever get the chance to work in true dialogue, you'll never forget it. You'll spend the rest of your career trying to recreate even a piece of that magic.

Okay, so now that you've mastered the skill of dialogue, you're all set, right? Not so fast, speedy. Let's delve into the souls of people a bit and explore another important spice for the management formulation.

From the Pep Talk to the PEP Talk: Motivating Without Melting Down

Before I had a team of my own, I was frequently on the receiving end of motivational messages from upper management. You know the ones: grand declarations about "transformation," "synergy," or "igniting purpose," usually delivered in a town hall with a slide deck full of vague Venn diagrams and stock photos of hikers on cliffs, or in a closed door conference room with a high-level executive whose bonus depended on our motivation to complete the projects in our queue.

I'd sit there thinking, *This is inspirational, the same way a rice cake is filling—technically it counts, but I'm still hungry.*

Then I became a manager... and the joke was on me.

It turns out motivating a team is a bit like making an emulsion. On the surface, it looks easy. Just mix oil and water, right? Well, without the right emulsifier (in this case, trust, clarity, and just enough humor) and the exact amount of pressure, it separates faster than soda on a hot loading dock.

As a new leader, I thought motivation came from the perfect email or a heartfelt one-liner during a team meeting. Spoiler alert: It doesn't. You can't just sprinkle inspiration like powdered sugar on a tough workweek and hope it tastes sweet. Motivation is more like sourdough: It needs time, care, and frequent feeding.

I learned quickly that everyone's motivator is different. For some, it's recognition. For others, it's autonomy, impact, or even free snacks. For most scientists I worked with, it's not a corporate acronym. Telling someone to "embrace GRIT" is like microwaving a soufflé—sure, you can do it, but you'll ruin it in the process. GRIT stands for Goals, Reps, Iteration, Truth: set the aim, do the reps, iterate fast, and tell the truth with receipts.

Metaphors From the Frontlines of Management

- **Motivation is like carbonation:** Overdo it and you get bloated with hype. Underdo it and you've got flat soda... and a flat team.

- **Leadership is a spice rack:** The same ingredient can bring a dish to life or ruin it. Use empathy like garlic: early, often, but never too much at once.

- **Inspiration is a slow cooker, not an air fryer:** The best results come from consistent effort over time, not blasting people with heat and expecting gourmet results in 15 minutes.

At some point, I stopped trying to be a motivational speaker and started behaving like a motivational human. I listened more. I laughed with the team. I told them the truth, even when it wasn't wrapped in corporate glitter. Somehow, that's what worked.

If there's one thing Yelena and I have learned, it's this—and stop us if you've heard it before: Your team won't remember the slides or the acronyms. They'll remember how you made them feel: seen, supported, and occasionally, like they weren't the only ones going slightly bananas during Q4 chaos. How many times has this been said in motivational and leadership speeches across the globe? Though it isn't our original idea, we still needed to discover it for ourselves. Much like when we tell our kids not to run on a wet surface, only for them to discover for themselves why we said it (and we're the mean ones?). If all else fails, bring snacks. No one's ever been mad during a donut meeting.

How do you make the "right call" when the future won't sit still? The bigger mistake isn't failing to predict tomorrow; it's pretending prediction is the job. The real work is what you do today: consistent effort on the few variables you actually control. Do that long enough and outcomes start to look "predictable" in hindsight, not because you saw the future, but because you shaped it.

Dan Brown's latest Langdon novel, *The Secret of Secrets*, toys with this same itch. Central to the story is the character Katherine Solomon's noetic science, the idea that human consciousness might anticipate or influence outcomes. It's catnip because it mirrors how great decisions feel from the inside: You move with conviction before proof arrives. In the real world, controversial "presentiment" studies claim that people show tiny physiological shifts before random computer events, though admittedly, the replications are mixed and fiercely debated. What's useful isn't the controversy; it's the discipline: act where you have leverage, gather evidence, and adjust. That's how you make your own "luck."

There's a leadership trap adjacent to this: the Dunning-Kruger effect, often summarized as, "The less you know, the more you think you know." Overconfidence in unfamiliar domains turns "vision" into expensive rework. I've watched managers, especially in the supplement business, make sweeping calls on blending, particle size, or homogeneity and then blame "the formula" when execution was the issue. This is avoidable. Ask for help. Define decision rights. Hire people who know more than you and listen to them. Expertise isn't a threat—it's a shield.

Practical Rule of Thumb:

- Control what you can (goals, reps-data collection, cadence, quality bars).

- Measure what happens (receipts, not vibes).

- Adjust fast (iteration beats prediction).

- Defer to experts outside your lane.

Do that, and you won't need a prophecy. You'll have a process.

Now that we've navigated the soft skill arts, we'll predict your next question: *How do I get one of those elusive promotions?* Easy there, cowboy (or cowgirl). Promotions are as unpredictable as a handful of flavored jellybeans; you might get toasted marshmallow... or you might get earwax. Promotions do happen, and when they do, there's usually a formula behind them.

José: Here's one of my favorites. When I took over a new team of scientists, one member confided that they had never been promoted in their entire time at the company. The feedback I'd heard from others was glowing: great teammate, strong skills, and solid reputation, but they just needed more exposure to senior leadership. That was an easy softball pitch down the middle. I asked them to set up a tasting and presentation for upper management, handed them the opportunity to run it, and stepped back.

Did they crush it? Absolutely. Did they get the promotion? You bet. Later, they asked, "Why me? What did I do differently to finally get promoted?" My answer was as vague as it was true: "You did it yourself. I just gave you the platform. You could have stumbled (and learned), and we would have tried again later, but you were ready, and you nailed it. Congrats."

Here's the thing: The skills that set you up for that promotion—building trust, delivering results, knowing when to shine and when to listen—are the same skills that will keep you relevant in the next era of work. Like it or not, the workplace is evolving again.

We've gone from filing cabinets to cloud drives, from landlines to video calls, from paper memos to Slack emojis. Now, there's a new coworker joining the mix, one that never takes a sick day, never needs a coffee break, and doesn't ask for a promotion: artificial intelligence.

Before you roll your eyes and say, "Great, José, just what I needed—a robot breathing down my neck," hear me out. AI isn't here to steal your job... at least not if you know how to work with it instead of against it. Believe me, the people who master that balance will have an edge as sharp as a chef's knife fresh off the honing steel.

Chapter Fifteen

Bittersweet: Food Science in the Age of AI and COVID

From Artificial Sweeteners to Artificial Intelligence

A rtificial intelligence is being called the latest "threat" to our industry, the lurking robot in the room. Many fear it will replace beloved roles, flatten career paths, and make some of our hard-earned skills obsolete. Honestly? That fear is understandable. New technology always comes with a shadow,

By focusing only on what AI might take away, we risk missing what it's already offering. Today's AI can help catapult your productivity in ways we haven't seen since the inception of the internet.

Need an email proofread and polished in seconds? Done. Need a literature review that used to take you six weeks? Now it's a weekend project. Need to brainstorm new product concepts while also mapping regulatory hurdles and financial implications? AI can do that before you've finished your coffee.

We can't predict the future, so we won't be able to verify that AI won't someday do more harm than good. Yet at this moment, it's an accelerant—and in scientific terms, a catalyst. It's an extra set of hands, an always-awake mentor, and a coach that never forgets what you told it last week. For us, AI has already helped register two businesses and assisted with the writing of this very book. Work that would have taken months, even years, was compressed into weeks.

In the past, if you wanted to start a company or write a book, you needed a human mentor who had done it before, someone whose time was expensive and limited. Today, we have our own "Jarvis" in our pockets, 24/7. This AI mentor has degrees in everything, can quote laws, and can walk us through incorporation paperwork as easily as a food scientist stirs sugar into water. Why not embrace it? Why not test-drive what it can do while it's here in this golden age? Take it for a spin. Your BRQ score will survive, I promise.

How AI Is Changing Food Science

In food science, AI is much more than a tool; it's quietly becoming a co-pilot. It's scanning white space in the market to identify unmet consumer needs. It's shrinking months of lab work into days. It's screening millions of probiotic strains before you've even stepped into a lab, which is now ubiquitously called "high throughput screening." It's writing our meeting summaries and organizing our research notes. AI doesn't just suggest flavors; it predicts how they'll interact in a matrix. Imagine knowing before you even start bench work whether your vanilla will mute the acidity in your strawberry base.

For the most part, the industry is beginning to embrace it, though we're still only scratching the surface. The fear of being replaced is still strong, but so is the opportunity to use AI to do more—faster and smarter than we've ever been able to before.

Pro tip: Embrace the unknown. The future will come whether you hide from it or not, so you might as well meet it with curiosity, not fear.

COVID

When I accepted the role in New Hampshire during COVID, the general belief was that the pandemic was a strange, temporary disruption. Yelena and I thought that within a few months, we'd be back in offices, kids would be back in classrooms, and life would snap back into place. So, we made what felt like a calculated gamble: We bit the bullet and moved up north, aiming to get the kids settled in time to start school that September.

School did start—masks, distance, and all—but my work reality was a different story. I ended up staying hybrid for almost my entire five years in the role. The office became less of a daily destination and more of a rotation, while home was transformed into mission control for both my work and my family.

Windham became our breather. The move gave us something we never truly had before: space. Space to breathe, space to spread out, and space to rethink how we wanted to live. And it wasn't just the physical space; it was the community. Neighbors waved from driveways; kids gathered in backyards, at the park, or in the jump houses and movie theaters a few towns over; and there was an unspoken understanding that we were all weathering this strange chapter of history together. Many times, help arrived before you even had to ask. Recall when our fourth child decided to arrive on the highway? A neighbor was at our door within minutes to watch the sleeping siblings. That's Windham. It's a place where roots grow deep, and where we started learning to balance comfort with ambition.

If you think back, you'll remember, COVID wasn't just a chapter. It was a jolt. One day, the world was open, and the next it felt like someone had flipped a switch and locked the door. Schools went dark, grocery store shelves went bare, and "normal" became a word we weren't sure we'd ever get back.

When COVID hit, Yelena was working in a flavor lab in New Jersey. Her role was deemed "critical," which meant that while many others in the corporate world set up makeshift offices at their kitchen tables, she was still commuting in, masked and gloved, to keep flavor development moving for customers who still needed product.

In those early months, we didn't know if this was going to be a blip or a long-haul battle. Like many families, we thought we should give it a few months, and then we'd all be back to normal. When fall came and schools were set to reopen, we made the call: We're moving. Our BRQ score increased that day since Yelena had to find a new company (she did, and it was a juicy choice), and we left the comfort of our community to join our new community up north.

The question for our industry was simple: How do we keep the food supply safe in infectious times? The answer was the same one we'd lived by for years: good manufacturing processes (GMPs). During COVID, we treated them like oxygen: controlled entry, obsessive handwashing and sanitizing, masks in the lab, staggered shifts, tight visitor rules, and contactless deliveries. Everyone who crossed a threshold wasn't just protecting themselves; they were protecting the food millions relied on.

Yelena: March 2020 flipped the switch. School and daycare closed overnight, and two little ones were suddenly home. We were lucky to have a teacher from daycare come to our house. That meant I could take the early lab shift (7 a.m. to 1 p.m.) while José covered mornings. Even with help, we lived on high alert, and it was a strange reality for a product developer who couldn't taste with peers. We had to invent new ways to work and new ways to take care of each other.

José: The real, daily conundrum wasn't Zoom; it was calendars versus kids. Two careers in one house meant constant triage: which meeting was truly career-limiting to miss, who needed to be on camera and who could dial in, who had a 10-minute flex window, and what we'd do if our sitter got sick or had to leave early. We built a quiet rhythm around that chaos, blocking the non-negotiables first, saying our hard stops out loud each morning, trading rescues so the load stayed fair, and defaulting to "video-critical stays, audio-optional flexes." When logic failed, we resorted to rock-paper-scissors. I lost more than I'll admit. That's how the house stayed mostly upright: not with perfect balance, but with fast, kind decisions and a running sense of humor.

If you've been there, you know that you also just did your best. Somehow, that was enough.

Chapter Sixteen

After Dinner Stories: Life Beyond the Lab

Throughout a long career, you tend to collect stories, the kind that make you think, "One day, I should write a book about this." Well, we're writing a book... so let's do just that.

The social side of corporate life often extends well beyond the lab bench or the office cubicle. It's where social bonds form and where coworkers get to see the other side of you—the one that exists outside spreadsheets, formulations, and endless email threads. Post-work activities are a common outlet: company softball games, pickup basketball, and the occasional after-work drink (purely for research purposes, of course—how else can we keep up with the latest trends in flavored spirits?).

Inside the workday, extracurriculars are just as common. Many companies have employee resource groups for shared interests, or host programs like Toastmasters to help people face the number one fear in America (no, not snakes—public speaking).

One of my favorite extracurriculars was indeed Toastmasters. I had a blast crafting speeches that landed exactly as awkwardly as you might imagine. Somehow, I made it into the deeper rounds of a regional competition, and my "success" became a talking point across the organization. At PepsiCo, we held quarterly town halls to recognize accomplishments, and these often included lighthearted videos. For mine, we spoofed Rocky, and for authenticity's sake, I even drank raw eggs like Stallone's character. The crowd's reaction was priceless and worth every queasy second. The photos floating around on social media... well, let's just say they're not my most flattering, but they're worth tracking down for the laugh.

PepsiCo was such a large company that it could field enough softball teams to run its own internal league and crown a champion each year. We'd even draw enough fans to fill a small bleacher section. People would bring snacks and drinks, and sometimes even fire up a grill. These games were actually networking opportunities disguised as fun. I played on the beach volleyball team and even joined a local town basketball team that Pepsi helped sponsor.

If your company offers activities like these, join in. They can be incredible for expanding your network, boosting your mental wellness, and creating shared experiences that strengthen working relationships. Just remember you're still representing yourself and your company. I've seen more than one career take a hit because someone forgot that "off the clock" doesn't mean "off the record."

The moments that shape your career and your connections aren't always in the meeting room. Sometimes they happen over a postgame beer, in a goofy town hall video, or during a laugh-filled lunch break. These are the spaces where trust deepens, relationships form, and people start to see the human being behind the title. It's yet another reminder that beyond the lab coats is where purpose lives.

Some of the most fulfilling extracurricular activities of my career were tied to the Adelante Employee Resource Group. I even led the organization nationwide for a year or so. It wasn't just fun; it gave me a way to make a real difference for the company outside of my day-to-day job.

Like many ERGs, we struggled at times to quantify our value to the business. We weren't just hosting cultural potlucks; we were surfacing insights into a consumer segment with purchasing power in the trillions. The projects that spun out of that work were as fascinating as the reasons some of them never launched.

I was once tasked with developing a heritage-inspired cola. Through Adelante, we helped translate cultural traditions into testable ideas. The concept advanced deep into internal evaluation before the business moved on. Even when ideas don't launch, the learning compounds, and the impact was real: Culture can drive concepts.

Adelante also took me to the Los Angeles market for a guided tour and a visit with the famed Mr. Cartoon. It opened my eyes to how deeply culture shapes a market's product mix. That wasn't knowledge I could have picked up in the lab.

One of my closest collaborators during that time was Lance, the "handsome" sales guy from Texas who, for reasons I still don't fully understand, enjoys calling me the "nerdy scientist from Valhalla." We worked together often, became friends, and kept that connection alive well beyond PepsiCo. Yelena also forged a strong friendship with Lance, and hers is marked by warmth, good conversation, and the occasional zinger when the moment's right.

You know, he'll tell anyone who'll listen that he was my "best man" at our wedding. Technically? No. He was invited, though, which he insists is basically the same thing.

Lance's Perspective

I still remember when Employee Resource Groups at PepsiCo were mostly social spaces, communities where you connected with people who shared your background or experiences. That sense of belonging mattered, especially in a corporate environment where you could easily feel like an outlier.

That's where I met José. How else would a handsome Frito-Lay sales rep from Texas cross paths with a nerdy beverage scientist in Valhalla, NY? Adelante pulled us together and pulled in members from across PepsiCo: Frito-Lay, Pepsi, Quaker, Tropicana, and Gatorade.

On a business level, it gave us a window into PepsiCo far beyond our own silos. On a personal level, it created bonds that last to this day. Yes, I was "basically" the best man at his wedding. (Look, I was invited. That counts.)

At first, the business connection of ERGs was... fuzzy. Like trying to connect a flip phone to Wi-Fi: technically possible, but no one knew how. We knew these groups mattered, but tying them to performance, strategy, and innovation wasn't straightforward.

That's why the shift from ERG to BRG (Business Resource Group) was so powerful. It wasn't just a rebrand; it was a declaration of purpose. Suddenly, we weren't just communities—we were contributors.

We influenced hiring pipelines, shaped product perspectives, and challenged assumptions. Leading a BRG became a proving ground. We ran meetings, managed budgets, and pitched to executives. Those weren't "soft skills"; they were leadership muscles our day jobs didn't always let us flex.

We had wins. Real ones. Millions in new business from ideas like Sabritas/Lay's Que Rico Chips and Pepsi Latino Flavors, which were born in Adelante focus groups. We saw them on store shelves, and other ERGs started doing the same.

We also had inspiration from our own "Godfather," Richard Montañez—yes, *that* Richard Montañez, of Flamin' Hot fame. His journey showed us that belonging and business impact can, and should, coexist. Doing that alongside my favorite nerdy scientist and his equally brilliant wife? That's the stuff you can't plan for, but it's the good stuff.

Trade Spend to Headspace

José: Lance could probably keep telling stories until we had a whole other book, and knowing him, he'd insist on calling it *Better Looking in Texas: The Lance Chronicles*, but this isn't his memoir (no matter how much he'll claim royalties when this book sells). So, let's pull the mic back, set him gently in the corner with his "Most Handsome ERG Member" sash, and return to our regularly scheduled programming. *¡Gracias, hermano!* Keep those jokes coming at my expense.

My time with Adelante taught me that work doesn't have to stay neatly inside your job description to matter. Sometimes the most meaningful contributions happen outside the cubicle walls, when you step into projects and experiences that connect people, culture, and purpose. Those moments reminded me that I could use my career as a platform for something bigger than the next product launch. As impactful as Adelante was, another experience took that idea even further. It didn't just broaden my perspective, but rewired it entirely.

One of the most profound extracurricular experiences of my career came when I had the rare chance to join Vitamin Angels™, a non-profit whose mission is to end malnutrition across the globe. Their work is the kind of thing that restores your faith in humanity and also reminds you how much work there still is to do.

In 2017, I traveled with their team to Roatán, Honduras. We visited communities scattered across the island, delivering vitamins and deworming medication to children and expectant mothers. It sounds noble on paper, but nothing prepares you for what you actually see. We walked down dirt roads littered with animal and human waste, surrounded by music blaring from tin-roof homes while curious neighbors leaned out of windows to watch us pass. The soundtrack was equal parts laughter, shouting, and the occasional barking dog.

The health conditions we encountered were jarring. Diseases considered rare or eradicated in the US were common here. Vitamin A and D deficiencies have led to preventable blindness in children. Malnutrition had etched itself into faces and postures, stunting not only physical growth but opportunity. Yet, there was resilience—kids smiling through their symptoms, parents welcoming us into their homes, and communities showing warmth despite hardship.

I came back changed. I realized that the starting line in life is not the same for everyone. I'm fine with the idea that we don't all start at the same point—that's life. However, in these communities, the starting line is so far back it's like they're being asked to run a marathon barefoot while everyone else gets a head start and a pair of Nikes. The heartbreaking part is that we know how to fix this. Food science has solved malnutrition in countless ways—fortified foods, supplementation programs, and micronutrient blends—yet millions still go without.

I may never cure cancer like my high-school classmates once imagined, but I can make one promise: Malnutrition, consider this your eviction notice. I'm formulating your demise, one micronutrient at a time.

Then there was Yelena's side of the story. While I was walking dirt roads in Honduras, she was navigating an entirely different mission: keeping our household from collapsing while solo-parenting a toddler. This was our unofficial pattern: one of us in the field chasing big career moments, the other holding down the fort at home.

Her perspective mirrored the one she had during the CBD project: pride in the work, admiration for the cause, and a deep awareness that this was exactly why we had chosen careers in food science. But also? A realistic acknowledgment that someone had to manage bedtimes, meal prep, and an active kid with a will of iron, while I was out saving the world, one vitamin at a time. She likes to joke that for every child in Honduras I helped that week, she rescued our own child from climbing on furniture or feeding Play-Doh to the cat. Wait, we didn't have a cat... whose cat were we feeding Play-Doh to?!

Our story is a reminder that behind every mission trip, launch, or milestone, there's always someone back home making it possible, and often, they don't get the spotlight. In our case, I'm just grateful that someone is Yelena. Because if there's one thing I've learned, whether in a remote Honduran village or our own living room, it's that feeding and protecting people is a team sport.

Rewind to 2013, the summer "Get Lucky" was stuck in everyone's head, and we were about to bet on our own luck. Different city, different stakes. We didn't have the Honduras perspective yet; what we had was a hunch, some courage, and a resignation letter.

José: The morning we resigned felt like stepping onto a moving walkway. We'd accepted offers with another company in our industry, and, pure coincidence or not, industry headlines lit up about a new strategic partnership between our future employer and a competitor of our current one. Either way, once you're headed to a direct competitor, many companies end access immediately. That's what happened: standard protocol, escorts included. The doors clicked behind us and, for a second, it felt like the gates of Valhalla had swung shut. Turns out, that's how new quests start; one hall closes, another opens.

How we got there matters. We weren't itching to leave. We had a good thing going: friends, rhythm, the comfort of knowing exactly where the good coffee was. Then that restless "grow" signal wouldn't shut up. We tossed in an application, half-expecting nothing. When interviews turned into offers, on terms we didn't think were possible, we looked at each other and said two things:

"We should've asked for more," and "Well... guess we're moving to Vermont."

Yelena: Our last day was cinematic. We'd just flown back from a family wedding overseas when messages started pinging: "Is it true?" The industry is small, so news travels. We decided to be direct: walk in, resign, and start the next chapter.

I dressed like I might be photographed: black suede heels with gold tips, tuxedo-style pants, a white shirt, and a blazer. Not my usual lab bench look, but I had a hunch I'd be walking out as soon as I walked in. Our campus had two buildings; José dropped me at mine and headed uphill to his.

I found my manager, handed over my letter, and before I could sit, we were already on the move to HR. No desk-sweep, no long goodbyes. If you've ever joined a competitor, you know the drill: access off, badges in, quick exit briefing.

Here's the part I didn't expect: the lobby couch. Our building had a front lobby where everyone passed through to get anywhere—labs, meeting rooms, you name it. I was asked to wait there while logistics were handled. Which meant every colleague, on their way to do anything, walked by... and waved... and stopped. It turned into a spontaneous receiving line. We laughed, traded stories, answered the "Vermont?" question a hundred times. It was weirdly wonderful. That lobby became our own little Valhalla, with friends filing past like a hall of heroes, offering blessings, jokes, and a hundred versions of "Vermont, really?"

Meanwhile, José was in a long project handoff with his manager. Hours later, he and his manager came through the lobby, headed to lunch. I waved like I was at a parade. HR rules meant I stayed put—no cafeteria cameo for me, so the lobby show went on. When

José was finally released, we walked to the car together, compared notes, and realized we'd had the same day in different movies.

José: Then we drove north. Our two years in Vermont were fast and formative. Year One, the place buzzed; new devices were on everyone's lips and whiteboards. Year Two, some bets didn't land, calendars opened up, and we learned a different skill: keeping momentum when the spotlight shifts. It was exactly the education we didn't know we needed.

Let's detour into New York City for a fun story before we continue.

Sampling, Stand-Up, and...Trump? A Day in Corporate Promotions

At some point in my corporate career, someone thought it would be a great idea to send the entire workforce to New York City for a massive beverage giveaway. Yes, nothing screams "efficient use of human capital" like deploying chemists, marketers, and finance managers to hand out icy cans to skeptical New Yorkers on a humid summer day.

If that's not already the setup for a comedy sketch, it gets better.

We traveled down by bus, and in a gesture of corporate hospitality, someone had the idea to entertain us with a comedy special by the performer we'd be seeing live after the day of beverage heroism. No one told the comedian this. So, after sweating our branded polos off and charming thousands of thirsty pedestrians, we sat down for the grand finale... and got word-for-word the exact same set we'd watched six hours earlier on a 13-inch bus TV.

To make things worse, in the middle of his routine, he drops a joke about a competing brand. You've never seen a group of loyal employees turn on a comedian so fast. The crowd booed, not aggressively, but with the kind of disappointed sighs you hear when someone ruins the office microwave. I felt awful for the guy. His face looked like he'd just lost the Super Bowl and stubbed his toe at the same time.

Now, you'd think we'd learn our lesson, but this is corporate America, where the postmortem often reads, "Let's do it again, but different."

So, the next year, after another day of beverage giveaways to the fine people of New York (who were slightly more receptive the second time around), we skipped the comedian. This time, we were treated to a motivational speech by none other than... Donald Trump. Yes, Donald Trump. Pre-presidency and pre-political firestorm, but still very much Donald Trump. What did he say? I have no idea. I think I blacked out somewhere between "You're gonna love this company" and "I only drink water when it's got my name on it." Truthfully, it was reminiscent and similar to his days on *The Apprentice*. Aspirational and admirable.

What did I learn from all this? Honestly... not much. Maybe that New Yorkers will always surprise you, comedians deserve empathy, and sometimes the best stories come from days where no one knows what's going on. Still, those were fun days.

Before we wrap up, let's talk about some of the moments that prove a career in food science is about far more than formulations and stability pulls. Sometimes, it's the unscripted experiences that stick with you—the kind you can't plan for, but wouldn't trade for anything.

Take the time we met Jimmy Buffett.

One of the wildest, most surreal encounters of my career didn't happen in a lab, a boardroom, or even at a corporate event. It happened in an elevator at the Hilton rooftop in Chicago.

We were in Chicago for the IFT Conference, one of the three big industry events. The kind of annual industry gathering where you network until your voice is gone and where the real magic happens after hours. After a long day of sessions, about 10 of us from food and beverage crammed into the elevator to head to the Hilton rooftop. An older gentleman, friendly and disarming, stepped in with us. He asked casually, "So, what do you all do?"

We explained that we were in the food industry, which was followed by some small talk, nothing remarkable... until someone in our group froze mid-sentence. You could see it dawn on them—that voice. That easy, island-lilt drawl. Could it be?

It was. Jimmy Buffett. Not Warren (no financial advice was dispensed that night), but Mr. Margaritaville himself.

Here's where it gets truly unbelievable: Instead of offering a polite nod and moving along, Jimmy decided to join us. Not for one drink. Not for thirty minutes. For four hours!

If you're wondering if we interrogated him like amateur detectives, you'd be right. We peppered him with questions, half-expecting to find some gap in the story that would prove he was just a very convincing impersonator. Nope. Every detail was spot-on, from *Fins Up* anecdotes to behind-the-scenes stories about his current tour.

Then came the clincher. He went to grab his Apple device to play us unreleased tracks from his upcoming album. There we were, a bunch of food scientists, sales reps, and marketers, dancing on a

Chicago rooftop to songs the world hadn't heard yet, songs played for us by the man who wrote them.

What struck me most was his humility. For four hours, he was just Jimmy, hanging out with strangers, drinking four glasses of decent-but-hardly-fancy rooftop-bar wine (maybe $45 total, $20 tops anywhere else). He signed autographs without hesitation. He posed for photos, including the now-infamous one of him dancing with Yelena that lit up our social media feeds for weeks.

The next morning, we were all checking our phones, laughing at how our industry friends, and even some family, were losing their minds in the comments (and texts). "How in the world...?!" "Only you two..." My favorite: "Was he wearing flip-flops?" (Yes. Obviously.)

When the Hilton finally shut down the rooftop for the night, Jimmy's shirt pocket was stuffed with business cards from our group. I'm sure most went straight into the recycling bin, but for those four hours, we were all living in Margaritaville, and Jimmy Buffett was the guy buying the next round. Well, technically, we did the buying because Jimmy's money is no good with us.

That happened a couple of years after another once-in-a-lifetime adventure, the kind you don't plan for, but when it happens, you just have to say yes. Case in point: the time I won Super Bowl tickets.

In 2010, early in our relationship, I won two tickets to Super Bowl XLV (played in 2011) in a PepsiCo raffle. That alone was surreal enough, but there was a catch: Yelena and I had only just started dating, and we weren't exactly broadcasting our relationship at work. PepsiCo was like a small town with vending machines, and news traveled fast.

The game was in Dallas, Yelena's favorite team's home turf. While the tickets were free, everything else about going to the Super Bowl was decidedly not. Flights, food, transportation—it all adds up fast when you're early in your career and very much making ends meet. Luckily, we had good friends from PepsiCo who had moved to Dallas and were kind enough to let us crash at their place. They were gracious, but also clearly intrigued (read: amused) at who I was bringing along. We all rolled with it.

PepsiCo made sure the whole trip felt like an event. We went to a private concert featuring some of the hottest artists of the time (we're pretty sure Jason Derulo was one of them), and a Q&A panel with the Manning clan—Eli, Peyton, and Archie—hosted alongside PepsiCo's CEO, Indra Nooyi. Harrison Ford, Terry Bradshaw, and other famous faces wandered around as if we'd all just happened to end up in the same football-themed universe. As a thank-you to our hosts, we joined them at the NFL Experience, a kind of football amusement park, complete with games, exhibits, and enough merch to fill an entire carry-on.

Game day arrived, and the excitement was electric. That is... until we found our seats. We knew they wouldn't be on the 50-yard line, but nothing could have prepared us for the nosebleed-plus section. We climbed all the way to the top of the stadium, and then discovered our section was above that. To get there, we had to squeeze past fans in the very last row, duck through a narrow passage, and emerge into a tiny, even higher section with maybe a dozen seats.

Let's just say that "watching the game" was more accurately described as "watching Lego-sized players run around while relying entirely on the Jumbotron for any real idea of what was happening." We didn't care, though—it was the Super Bowl. We were there, in the "Ice Bowl" (as it was dubbed for its unusually frigid Texas weather), bundled up and loving every second.

In true Super Bowl fashion, there was even a little drama. Christina Aguilera famously flubbed the lyrics to the national anthem that year. We didn't notice at all, maybe because of our distance, or maybe because we were too busy taking in the spectacle and only found out later when it dominated the news cycle.

By the time the game wrapped up, our cheeks hurt from smiling. It was one of those surreal, pinch-me experiences that we still laugh about. Now, once we got back, the questions from coworkers started almost immediately: "You took... who?" "Were you two...?" Let's just say our "unofficial" status didn't stay unofficial for long after that.

Then there were Yelena's truly memorable first two weeks at work. First week: a softball game. One swing, one bad bounce, hand injury number one. Second week: a company basketball game. She's wide open downcourt and catches a perfect pass... with her thumb. It bent in a way that made even the toughest teammates wince. Hospital visit number two. Same month. Same new job. I felt like a corporate knight in shining armor picking her up from the hospital (after I finished the game, of course), though I'm not sure she recalls my heroism the same way.

Yelena: I had to find my way to the hospital myself. I was in pain, and that was when I needed you, not afterwards when I was on painkillers. Thanks, anyway.

José: Oh, and the company briefly considered banning sports altogether.

Individually, these stories are fun. Together, they're proof of something important: This industry runs on relationships, serendipity, and the connections you make in between the "official" work. The people you bump into at conferences, the co-workers who share those ridiculous once-in-a-lifetime moments, the mentors you never expected to meet—these are the threads that weave your career together.

Here's the thing: In this industry, the more years you're in it, the more these connections and shared stories matter. Which brings us to something we've learned over and over again...

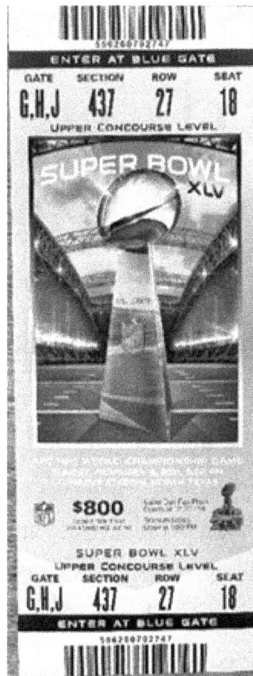

Figure 3. "Once-in-a-lifetime seats... in the section above the nosebleeds. Still worth every frozen minute."

Chapter Seventeen

The To-Go Box: Lessons to Carry Forward

By now, you've already heard me talk about my grandmother's house, the metal gates, the comforting chaos of street noise, and the surprise bite of lime peel in her oatmeal. I'll spare you the full replay, but I bring it up again for one reason: Those little moments shaped how I see people, work, and life. They taught me that comfort and challenge often live side by side, and that it's the mix of the two that makes things meaningful.

That lesson plays out constantly in our industry. Both the dietary supplement and food and beverage worlds are small enough that your paths will cross with the same people again and again. Sometimes you'll celebrate together. Sometimes you'll clash. Sometimes you'll chew on something unexpectedly bitter. Through all of it, the relationships you build, especially with mentors, are what make the difference between just surviving and thriving.

Mentorship isn't a luxury; it's a lifeline. In my experience, it's rarely formal. The best mentors tend to appear naturally, often in the moments you least expect. They might be someone a few levels above you, a peer who sees your potential, or even someone outside your function entirely. They offer you more than advice; they give you perspective, confidence, and, most importantly, trust.

For those familiar with Patrick Lencioni's *The Five Dysfunctions of a Team*, you already know that trust is the foundation on which everything else is built. Without it, feedback becomes guarded, collaboration turns transactional, and progress slows to a crawl. A good mentor cuts through that. They trust you enough to be candid. You trust them enough to listen. That's where growth happens.

Don't limit your idea of mentorship to people you've met in person. Some of our biggest influences have been authors, musicians, and public figures we've never met. A book that challenges your thinking, a song lyric that lands in exactly the right way, a story that reframes your own experience—all of these can be mentors in disguise. Inspiration doesn't care whether it shakes your hand first. The key is to stay open to it. Because when you do, you start to realize that life, like those mornings at my grandmother's table, will always be a mix of warm comfort and sharp surprises. The right mentors, in whatever form they take, help you savor both.

Salsa Time

Yelena: There was this one time, back in our pre-marriage "let's-take-the-red-eye-and-figure-it-out-later" days, when we were flying home from Puerto Rico. We landed groggy and running on zero sleep, only to remember that I had a salsa competition that morning (the food kind, not the twirl-your-hips kind.) I had just a couple of hours from the moment our wheels hit the runway to get ready for a campus-wide taste-off. The winner would earn a trip to Frito-Lay headquarters in Texas to compete against their best.

Luckily, all my ingredients were already in the fridge. The catch? My recipe for *Summer* "Salsa With Grilled Corn" called for charred green onion and corn. So, there I was, at the crack of dawn, firing up the grill while the neighbors in our apartment complex were heading out for their morning workouts. I pulled it together, showered, sprinted to work, and somehow made it just in time for the tasting. By the end of the day, the results were in: I had made the final four. Plano, Texas, here I come.

The finals happened to coincide with the kickoff to Hispanic Heritage Month. I started the morning with *dulce de leche*–filled churros (pure joy), then spent hours with Frito-Lay's top executive chefs touring their kitchens, a dream for any food scientist. Lunchtime was game time. The final four salsas from across the company were lined up. Presentations began, and when they announced my name, flashed my picture, and crowned my "Summer Salsa With Grilled Corn" as the winner... it was surreal.

The prize? An iPad, which in those days was a very big deal. But the *real* win was this: I was still a brand-new scientist, my rookie badge practically shining, and I had just scored a career highlight that would stay with me forever.

Looking back now, this wasn't just about salsa. It was a lesson in the ingredients that matter early in anyone's career: passion, energy, and a pure, unfiltered love for what you do. Those are powerful

forces, but they work best when paired with experience and ma-
turity. As the years go on, it's easy to let the fire fade under layers
of process, politics, and "practical thinking." The real challenge is
keeping that rookie hunger alive—the "eye of the tiger"—no matter
how seasoned you become. That's something worth taking home in
your to-go box.

Breadcrumbs of Philosophy: Where I Look When I Need to Look Up

José: If you've made it this far in the book, you've probably realized
we don't have all the answers. (Surprise!) We have picked up a few
universal truths that have guided us, sometimes stumbling upon
them in the middle of a lab experiment gone wrong, and sometimes
while wrangling kids who believe 9 p.m. is a reasonable time to start
building a Lego city.

Some of these truths we've discovered the hard way. Others were
passed down through mentors, books, or the occasional refrigerator
magnet. But if you're looking for the roots of the mindset that got
us here, and helped us stay here without completely losing it, two
sources come to mind:

1. "If" by Rudyard Kipling

This poem is like a resilience blueprint dressed up as verse. It
doesn't just tell you to be strong; it tells you *how*. How to keep
your head when everyone around you is losing theirs (i.e., Monday
morning status meetings). How to trust yourself when others doubt
you, but also make space to question yourself when needed (hello,
imposter syndrome). How to meet Triumph and Disaster and treat

those two impostors just the same... which, let's be honest, sounds a lot like back-to-back quarterly business reviews (QBRs).

It's aspirational, yes, but also incredibly grounding. I've returned to this poem more than once when my career (or my confidence) was wobbling.

2. The Writings of Yuval Noah Harari

While I don't share his existential panic over AI (you're doing great, ChatGPT), Harari's reflections on the evolution of humanity, belief systems, and social constructs helped me zoom out. Way out. In a corporate world where you can spend three weeks debating font size on a launch deck, Harari reminds us of the bigger picture. His books *Sapiens* and *Homo Deus* helped me connect the dots between the biology of hunger, the psychology of teamwork, and the illusion of certainty that permeates everything from grocery aisles to shareholder meetings.

Harari doesn't give you a to-do list. He gives you perspective. Sometimes that's exactly what you need when everything feels like it's on fire, again.

Fuel for Your Fire

I'm not saying these will be your guiding lights. Maybe yours is a song lyric, a parent's advice, or something a stranger once said that never left you. If you're looking for places to start, Kipling and Harari are pretty good traveling companions.

Fair warning: Neither of them will hand out stress balls, but they might help you handle the stress better. Speaking of stress, before we move on, there's one more truth you'll want in your back pocket.

If you're new to corporate America, or thinking about joining soon, know this: Not everyone will want to see you succeed. Egos abound, and if you don't learn to navigate them, they can quickly drain the joy out of your work. Your best defense? Find teammates who truly cheer you on and who will back you up when you need it most. When you find them, return the favor.

"Fire in the belly" is a phrase many successful people use to explain how they've made it so far. That grit, that unshakable drive, is the secret sauce to a career defined by continual growth. If you take only one thing from this book, it's to always listen to your wife. But if you take two, it's to find a way to light that fire in the belly because that's what moves you toward what might look impossible... or as scientists like to call it, improbable.

Michael Jordan, the greatest basketball player of all time (sorry, Gen Z), used to invent reasons to get worked up before a game so he could use the emotion to fuel his performance. It wasn't ego. It was ignition. That ignition turned him into poetry in motion.

I've always believed "all good things come to an end." After seven years in the international department, I moved to another role, and the real lessons in that chapter weren't technical but political. That shift taught me that change isn't scary. In fact, it should be embraced, or even sought out. My old phrase was incomplete: If all good things come to an end, it's because something even better is about to start.

If you've ever wondered what "fire in the belly" feels like, pause here. Sit somewhere quiet and queue up *From Within* by Mark Petrie. Let that slow build wash over you—the pulse, the swell, the surge. *That's* the feeling. That's the soundtrack to ambition, curiosity, and the refusal to settle. It even works while tackling that mountain of dirty dishes in your sink (I don't think I didn't notice).

Arnold Schwarzenegger wrote a book titled *Be Useful*. Barack Obama once said something along the lines of, "Get things done and people will notice." The ability to be productive could have its own scale, similar to the intelligence quotient (IQ) or the emotional quotient (EQ). If you get things done in corporate America, you will succeed, and people will notice. If you can combine productivity with prioritization, then you've unlocked a major secret for growth and advancement.

Here's the part that doesn't get talked about enough: Doing great work isn't a solo act. The most meaningful wins, the biggest leaps forward, the moments where the "fire in the belly" turns into something tangible—those almost always happen when you have the right people beside you.

For me, the person beside me was Yelena. Always. In every chapter of my career and our life together, we've been co-architects of the plan, co-navigators of the detours, and co-conspirators in the moments of creative chaos. We don't just form a partnership; we form an operating system. One we've refined, stress-tested, and run every single day without fail.

Problems can seem insurmountable, both in your career and in life. How have we tackled problems and solved mysteries? While there is no one single way or one single secret, there are some tried-and-true ways to make large mountains seem like molehills. One such way is by watching the movie *The Martian* with Matt Damon. Yes, a movie can help you with life if you apply the right lens. In it, Matt Damon's character faces what seem like impossible situations and handles each by breaking it down: solve the most important problem in front of you, then move to the next.

How about our kids? We used to believe that the ultimate parenting goal was comfort. A warm home. A fridge stocked with yogurt tubes and string cheese. Bedtime stories read with voices. The absence of fear.

And yet we know the truth, don't we?

Comfort rarely breeds resilience.

It soothes. It shelters. It whispers, "Why risk it?" But it rarely sparks the kind of internal fire that pushes you forward, that demands you try again, that teaches you who you are when the plan falls apart.

As parents, we wrestle with this daily. We want to give our kids a better launchpad than we had. But in doing so, we risk over-cushioning the ground they're meant to leap from.

We want to build them a safety net, but not a hammock. A launchpad, not a lounge chair.

The irony? Most of the things that built our own grit—starting over, moving often, awkward transitions, unmet expectations—are the very things we try to buffer our kids from. And we get it: Discomfort is uncomfortable, especially when it's your child squirming in it.

But when we think back to the hardest chapters of our lives—the ones we almost didn't survive, the ones that made us question everything—that's where the real growth lived. And often, it was the fire, not the blanket, that shaped us.

So, now we ask ourselves: How do we raise kids with a fire in their belly when the world hands them iPads and padded playgrounds?

Maybe it starts by letting them see our own fire flicker and flare. Maybe it starts by telling the truth about how clarity came late, how comfort can be a trap, and how fear sometimes wears the costume of practicality. Maybe it starts by choosing growth ourselves, even when it's messy.

Because the truth is, comfort is a great place to catch your breath. But it's a terrible place to build your life.

That's how we've done it, too, in our careers, in our lives together, and in building a family. One problem at a time. One shared goal at a time. The way we've been able to do that consistently, through all the chaos and change, is because we've built something between us that we trust completely.

The Barbosa Algorithm

Yelena's grandmother, a loving Russian *babushka* with a heart as big as her no-nonsense streak, once handed me a vodka shot for a cold. In her world, the cure wasn't in a medical journal; it was in tradition, family lore, and the unwavering conviction behind her eyes. I took the shot. Did it help? Hard to say. However, I felt the love, the legend, and the sheer will for it to work as it went down—smooth, bracing, and oddly comforting.

Growing up in Puerto Rico, my remedies were straightforward: Vicks VapoRub on my chest, plenty of water, and an early bedtime. No mystery, no ceremony, just eucalyptus fumes, hydration, and sleep. For the longest time, Yelena's *babushka* thought I was from Costa Rica, which I suppose only added to her curiosity about my "island" ways.

So, when I was dating Yelena and caught a cold, I figured I knew the drill. I didn't. Her *babushka* took one look at me, shook her head at my Caribbean remedies, and vanished into the kitchen. She returned with an unlabeled glass jar that had a large chunk of honeycomb floating inside, as it were guarding some ancient secret. She dusted off the lid; clearly, this wasn't from the liquor store but from her archives.

She poured the golden liquid into her fanciest shot glasses, one of the few treasures she'd brought from the USSR after it dissolved, and slid one toward me with the quiet authority of a woman who had cured more colds than I'd had birthdays. Could I question what was in it? Or the alcohol content? No chance. You don't question *babushka*. You just take the shot, feel the burn, and hope it cures more than the sniffles.

Moments like this remind me that science and data only get you so far. The rest comes from people, culture, and tradition, often delivered with the kind of conviction no peer-reviewed journal could ever match. That's the real lesson: Beyond the lab coats, formulas, and protocols, you sometimes find the very thing we're all looking for: purpose.

That's the essence of us. Our marriage is its own kind of folklore, written in moments big and small, backed by faith in each other's abilities and the unshakable belief that together, we'll find the cure for whatever "cold" life hands us.

No dual-voice memoir of a loving couple would be complete without an ode to each other, a tribute to the chemistry that fuels our life together. Ours isn't accidental. It's intentional. It's been built, tested, and strengthened over years of shared dreams, shared risks, and shared victories.

We call it *The Barbosa Algorithm*. It's the unspoken code that keeps us in sync. We make each other better, not by trying to change one another, but by helping each other grow into the truest versions of ourselves. There's no competition between us, no scorekeeping, no egos at war. We balance each other's styles, leaning on our differences as much as our similarities. When one of us is strong, the other can rest. When one of us is learning, the other teaches. Our pacing is naturally aligned, mine a little faster, but never so far apart that we lose step. We've learned when to step forward, when to step back, and how to protect the trust that sits at the heart of it all.

The words in our ketubah aren't just ceremonial. They are the original source code for the algorithm—the promise to walk life's path together, to encourage personal growth, to be honest, loyal, and devoted, to comfort each other in both joy and sorrow, and to create a home filled with love, peace, freedom, compassion, laughter, generosity, and respect.

It's not flawless. No algorithm is. Yet it works because we keep refining it, recalibrating with every challenge and every win. Like any great chemistry, the reaction between us keeps producing something worth sharing with the world.

Maybe that's what this whole journey has been about—not just building a career, a family, or a collection of stories, but building something that lasts because it's built together. A marriage. A family. A legacy. A shared dream that's flexible enough to bend with life's surprises, but strong enough to hold under the weight of everything that matters.

If there's any formula worth protecting, perfecting, and passing on, it's that one. It's our hope that in reading these pages, you've found not just stories, but sparks—reminders that even in an unpredictable, noisy, and sometimes overwhelming world, you can create something that's deeply your own.

So, keep experimenting. Keep learning. Keep refining your own algorithm. If you ever find yourself needing a little extra push, don't be afraid to call out for someone to open the gates. You never know what's waiting for you on the other side.

Conclusion: The Final Sip

S o, as the table clears and the stories end, we raise one last glass. Not of champagne, not of whiskey, but of ice wine—golden, sweet, and hard-won. Let's toast to a reminder that life, at its richest, is meant to be sipped slowly, with gratitude.

If you'd told me, back when I was scrubbing beakers on an eight-day contract, that one day I'd stand in boardrooms, lead global teams, taste lab-made proteins that smelled like boiled socks, and write a book about it all, I'd have laughed you out of the lab, after making you taste my latest formula. Yet here we are.

As I've recounted, when I was young, I thought I'd wear a white coat with an M.D. stitched over the pocket and cure cancer. Life, as it often does, laughed and redirected me toward food science. While I never became the doctor I imagined, I did get to work on something just as vital: nourishing people, preventing deficiencies, and extending health in ways that are quieter but no less important.

Still, I don't believe the story is finished. Maybe, years from now, I'll circle back, step into a classroom again, chase a late degree, or contribute in ways that surprise even me. The beauty of purpose is that it doesn't expire. It evolves.

Because that's the truth we've lived: Your journey is never done. No matter how old you are and no matter how many chapters you've already written, there is always room for another pivot, another pursuit, and another unexpected turn that adds meaning to the whole.

So, if one day you hear that I've gone back to chase that old dream, whether it's through science, advocacy, or a classroom desk, don't be surprised. Purpose isn't a finish line. It's a lifelong experiment. I'm still running tests. I tell my kids I want to be a marine biologist when I retire. We'll see!

We've covered a lot of ground together, from product launches to parenting, from gold standards to golden handcuffs. And while this book has zigzagged between food science, leadership, and life, one thread ties it all together: Careers, like life, are built on resilience.

Hemingway wrote, "The world breaks everyone, and afterward many are strong at the broken places." That's as true for a food scientist as for anyone else. We work in iterations: tweak, test, fail, adjust, repeat. Stability isn't found in never breaking; it's found in rebuilding stronger where the cracks once formed.

Think of it like stabilizing an emulsion. Sometimes the suspension breaks, the oil separates, and it looks like the whole thing is ruined. Nonetheless, with the right process, the right ingredients, and the patience to rework it, you can bring it back together, stronger, smoother, and more stable than before.

That's the real work of a career: not avoiding the breaks but using them to upgrade your formula. High BRQ scorers don't fear the fracture. They bend before they break, adapt before they burn out, and let every challenge refine them.

So, when the next challenge comes, and it will, remember: You're the product. Formulate yourself well. Test yourself often. When you break, rebuild in a way that leaves you unshakable.

It seems only fitting to mention Valhalla, the place where Yelena and I began at Pepsi's R&D labs. This town shares its name with the Norse Hall, where warriors go after the battle is done, and, incidentally, the cemetery across the road where many great innovators (and a few stubborn projects) have been laid to rest. In Norse legend, the gates of Valhalla opened to those who fought bravely. We like to think that the gates of career Valhalla opened, too, for those who dared to be curious, took risks, and kept laughing, even when the lab smelled like a chemistry experiment gone wrong. In the end, Valhalla was never a building. It was any place we earned by showing up for the hard thing, together. Gates don't close on you; they open when you outgrow the room.

Food science is iteration. So is life. You try, learn, adapt, and improve. Our hope isn't just that you learned a little more about food science; it's that you see your own story with fresh eyes. That you find your raw ingredients, keep tweaking, and stay open to the pivots that lead you somewhere unexpected.

While we didn't write this book to debate every piece of technology in our food system, I'll leave you with this: Innovation, whether it's GMO technology, AI-assisted science, or something we haven't yet imagined, is only as valuable as the purpose driving it. If that purpose is rooted in the benefit of all—in nourishing people, protecting our resources, and making life better for more of us—then it's a purpose worth pursuing.

We believe in a world where people bring their best selves to work, where safe, nourishing foods are within everyone's reach, where malnutrition is a thing of the past, and where everyone has a chance to succeed in this beautifully chaotic, unpredictable life.

So, stop taking yourself too seriously. Focus on what truly matters. Get yourself right, then turn around and help someone else do the same. That's a formula worth perfecting.

Scientists aren't always the best storytellers. We've tried our best here to change that, to turn the data, the projects, the failures, and the wins into something you could feel, not just understand. Whether we pulled it off is for you to decide.

We started as scientists. We learned to be leaders. Somewhere along the way, we became storytellers. If there's one truth we carry forward, it's this: Beyond the lab coats is purpose, and that purpose is yours to write. Because beyond the lab coats, beyond the formulas, and beyond the corporate meetings, purpose is what lasts. It's the ingredient that makes all the other ones matter, written not in pencil to be erased, but in indelible ink that time, failure, and success cannot wash away.

Benchside Ritual

Click. Pipette down. Drop in place.
Watch the titration find its pace.
Brix check—sweets on track,
Overage guarding the flavor's back.
Taste. Adjust. Taste again.
That's the loop from now 'til then.

From every trial, each careful pour,
We're chasing something worth much more.
Not just flavor that delights the tongue,
But strength for the old, hope for the young.
To fill the gaps where diets fail,
Sending good health in every pail.

From raw to refined, sip to sell,
We craft the stories your palate will tell.
Yet the quiet truth, the deeper art—
Is feeding the body, and healing the heart.

Further Reading

Institute of Food Technologists. *About Food Science and Technology*— a solid primer on what food science covers (perfect for non-scientist readers).

Gary Keller with **Jay Papasan.** *The ONE Thing* — a practical lens for focus and trade-offs we reference throughout the memoir.

Vickie Kloeris. *Space Bites* — behind-the-scenes of feeding astronauts; great "R&D meets constraints" stories.

Richard Montañez. *Flamin' Hot* — scrappy innovation and intrapreneurship inside big food.

Arnold Schwarzenegger. *Be Useful* — mindset and daily tools for momentum.

Perry M. Smith & **Jeffrey W. Foley.** *Rules & Tools for Leaders* — compact field manual for leading teams.

Andy Weir. *The Martian* — the most entertaining ode to problem-solving under pressure.

About the authors

José L. Barbosa & Yelena Barbosa are food scientists turned storytellers who love translating the language of the lab into something you can taste at the table. Over the past two decades they've worked across the food, beverage, and dietary supplement industries—building teams, launching products, and leading research from benchtop ideas to global brands.

José is a product developer and leadership-minded operator who has served as Head of R&D and Regulatory. He is a co-author on a scientific publication on iron absorption and holds an MBA in Leadership. Through Endless Thesis, he consults with companies on innovation, quality, and scale-up, and speaks on practical science, resilient teams, and the power of clear decision-making.

Yelena is a research and innovation leader with a gift for systems thinking—connecting sensory, quality, and consumer insights to create products that actually work in the real world. She mentors scientists and young leaders, champions kindness as a competitive advantage, and believes curiosity (plus a little humor) keeps any kitchen—from home to corporate—running.

Together they've twice worked side-by-side in the same company, proof that a shared mission beats separate calendars. They write to offer a behind-the-scenes pass to food science: the experiments that worked, the ones that didn't, and the choices that shaped a career and a family. When they're not testing recipes or reviewing formulas, you'll find them speaking at universities and industry events, consulting with mission-driven brands, or gathering friends around a too-small table.

Yelena and José L. Barbosa

Connect and learn more at: www.endlessthesis.com

www.ingramcontent.com/pod-product-compliance
Lightning Source LLC
Chambersburg PA
CBHW071555210326
41597CB00019B/3263